THE CLIMATE ADAPTATION GENERATION

A Blueprint for the Future

Robert W Collin

Copyright © 2025 Robert W. Collin

All rights reserved. No part of this publication may be reproduced, distributed, or transmitted in any form or by any means—electronic, mechanical, photocopying, recording, or otherwise—without the prior written permission of the publisher, except in the case of brief quotations embodied in critical articles and reviews, or as permitted by applicable law.

Library of Congress Cataloging-in-Publication Data is available.

ISBN 979-8-9993836-3-1 (paperback)

First Edition: July 2025

Publisher
Robert W Collin
Address
8388 Buckeye Court Frederick,
MD 21702

Cover design by Aly Sam (www.alysam.design)
Printed in the United States of America
eBook version is available for all major platforms

TABLE OF CONTENTS

Preface

Chapter One ----- 10
Adaptation and the Resilient Generation and Power

A. Demographic Influence and Numbers of the Adaptation Generation ----- 11
B. Demographic Influence and Power of the Resilient Generation
C. Environmental Context: Wastes and Disasters
D. Flourishing is the Goal for Everyone
E. The Role of Artificial Intelligence in Human Flourishing in a Changing Climate
F. Conclusion ----- 25

Advocacy Brief ----- 25
Planners Toolkit
Resources ----- 27

Chapter Two ----- 28
Emergency Responders and Climate Change

A. Overview ----- 29
B. Applications of AI in Emergency Responses to Climate Events C. Future Potential and Global Recommendations
D. A Case Study of Japan: A National Response
E. Conclusion ----- 43

Advocacy Brief ----- 43
Planners Toolkit
Resources ----- 50

Chapter Three ----- 51
Floods

A. Adaptation Strategies: Categories ----- 52
B. Examples from Around the World
C. Challenges to Adaptation

D. Future Directions
E. Adaptation to Flooding: Specific Strategies, Examples and Challenges
F. Global Focus: Flooding in the IPCC (Intergovernmental Panel on Climate Change) Sixth Assessment Report
G. The Role of Artificial Intelligence ------- 62

Advocacy Brief ------- 62
Planners Toolkit
Resources ------- 67

Chapter Four ------- 68

Storm Intensification

A. Climate Dynamics ------- 69
B. What Does Storm Intensification Look Like?
C. Examples of Storm Intensification
D. AI Examples Applied to Storm Intensification: How Artificial Intelligence is
    Shaping Climate Risk Awareness
E. Storm Track and Intensity Projection Maps ------- 77

Advocacy Brief ------- 78
Planners Toolkit
Resources ------- 84

Chapter Five ------- 85

Drought

A. Climate Context and Urgency ------- 86
B. Objectives of a Drought Adaptation Strategy
C. Frameworks for Implementation of Drought Adaptation Strategies
D. Continental-Scale Groundwater Flow: New Knowledge
E. Integrating AI and Groundwater Pathway Mapping in Drought Adaptation
F. Examples of Drought Management
G. Conclusion ------- 96

Advocacy Brief -- 96
Planners Toolkit
Resources --- 102

Chapter Six --- 103

Wildfires

A. What is a Wildfire -- 104
B. Examples of Wildfire Climate Adaptation
C. Use of AI in Wildfire Adaptation and Response ------------------------ 109

Advocacy Brief --- 109
Planners Toolkit
Resources -- 114

Chapter Seven --- 115

Earthquakes

A. What is an Earthquake? -- 116
B. Global Urbanization and Earthquakes: A Growing Collusion of Risk
C. Basic Climate Adaptation
D. Earthquakes and Climate Adaptation: Integrated Risk and Climate Adaptation for a Warming World
E. Integrated Approaches: Land Use, Construction and Nature Based Adaptation
F. Expanded Use of Artificial Intelligence for Earthquakes and Climate Adaptation
G. Expanded Case Studies of Earthquakes and Climate Adaptation
H. Conclusion: Towards Multi-Hazard, Equitable, AI Informed Climate Adaptation --- 125

Advocacy Brief --- 125
Planners Toolkit
Resources -- 129

Chapter Eight --- 130

Designing for Climate Adaptation: The Role of Place

A. Why Place Studies are Better for Climate Adaptation — 131
B. Designing for Adaptation I n Place C. Examples of Storm Intensification
C. Examples of Place Based Adaptation
D. Understanding Place Studies and Cumulative Impacts
E. AI and Place Based Planning — 140

Advocacy Brief — 141
Planners Toolkit
Resources — 143

Chapter Nine — 144

Climate Education Supporting Learning, Adaptation and Flourishing

A. The Education of the Climate Adaptation Generation — 145
B. Climate Anxiety
C. Confrontational Discussions in the Classroom: Balanced Messaging
D. AI in the Classroom
E. Conclusion — 149

Advocacy Brief — 150
Planners Toolkit
Resources — 153

Chapter Ten — 154

Climate Adaptation and the Health of the Climate Adaptation Generation

A. The Climate Change Generation and the Risks They Face — 155
B. Urbanization
C. Waste Heat
D. What Can Be Done to Address Urban Heat?
E. The Most Vulnerable are The First
F. Cumulative Impacts
G. Disease Expansion
H. The Diseases
I. AI and Disease: Examples
J. Conclusion: AI as a Critical Tool in Climate=Health Adaptation — 166

Advocacy Brief ------ 166
Planners Toolkit
Resources ------ 170

Chapter Eleven ------ 171

Blueprint for Flourishing

A. The Relationship Between Visioning and Flourishing in Climate Adaptation --- 172
B. Where to Start?
C. Legislative Frameworks Supporting Visioning and Blueprints for Flourishing
D. Conclusion ------ 177

Advocacy Brief ------ 177
Planners Toolkit
Resources ------ 180

Chapter Twelve ------ 181

Climate Change Impacts on the Natural Environment and Adaptation Measures

A. The Natural Environment and the Scale of Climate Change ------ 182
B. Jet stream Changes and Climate Changes: Causes, Impacts and Consequences
C. Impacts of Jet Stream Changes on Natural and Human Systems
D. Changes in Ocean Currents and Their Climate Impacts
E. Impacts of Changing Ocean Currents
F. Impacts on the Natural Environment
G. Extinctions Due to Climate Change
H. Case Studies of Climate – Linked Extinction
I. Ripple Effects: Ecosystem Collapse and Human Impacts
J. Adaptation Plans and Strategies
K. AI Applications to Protect the Natural Environment from Climate Change
L. Conclusion ------ 192

Advocacy Brief ------ 192
Planners Toolkit
Resources ------ 194

Chapter Thirteen — 195

The Climate Adaptation Revolution and Evolution

A. Is it climate Adaptation Revolution or Evolution? — 196
B. Climate adaptation and Capitalism: Tensions and Changes
C. Climate Adaptation and Marxism: Reclaiming the Economy in An Age of Crisis
D. Generational and Class Differences in AI
E. Digital Divide and Climate Adaption: Unequal Access in an era of Crisis
F. Conciusion — 207

Advocacy Brief — 207
Planners Toolkit
Resources — 209

Chapter Fourteen — 210

Climate Adaptation: Summary and a Blueprint

A. Our Journey Together — 211
B. Climate Migration: Breaking the Border Paradigm
C. The digital Divide and Power for Global Climate Adaptation
D. AI and Climate Adaption Futures
E. Cumulative Impacts of Climate Change
F. Conclusion — 224

Glossary — 225

PREFACE

This book is intended for educators, students, parents, grandparents, policymakers, urban and environmental planners, activists, advocates, and general readers. It is meant to help all readers be the change they want to see, to flourish in a time of escalating known and unknown climate impacts.

Here's a powerful quote from former President Barack Obama about climate change:

> *"We are the first generation to feel the impact of climate change and the last generation that can do something about it."*

He delivered this to both at the 2014 UN Climate Summit and later at COP21 in Paris, underscoring the urgent window of opportunity we now occupy. It is this sense of urgency that frames this book. The book is organized to be readable and implemented by many. Each chapter includes successful case studies, outlines of advocacy briefs, and implementable planning guidance, all of which apply AI. It concludes with a series of actionable next steps for the Climate Adaptation generation.

Chapter One

Adaptation and the Resilient Generation and Power

provides an overview of the adaptation generation and the resilient generation, including their demographics, power, and impacts from climate change. These are the people who can begin the adaptation process. What does it mean to be flourishing in the face of climate change? This visionary goal is attainable and is answered in this chapter. But how to get this going in real life, right now, is the big question.

Chapter Two

Emergency Responders and Climate Change

covers emergency responses to natural disasters. The role of Artificial Intelligence (AI) is discussed in detail with references to programs you can use right now. Prevention, preparation, and recovery are detailed. Case studies and examples of cutting-edge, successful approaches are fully explained. Advocacy briefs to legislators and courts are presented in useable detail. Land use is given special attention, with guidance on how to approach it that can be implemented immediately.

Chapter Three

Floods

focuses on floods. Floods are among the deadliest natural disasters. With storm intensification, discussed in the next chapter, floods can occur quickly and with little notice. Damage to the natural environment and property can be extensive and sometimes exceed the ability of insurers to cover. Floods can leave behind a wake of significant public health risks as standing water can create ideal breeding grounds for disease-carrying mosquitoes. Here, we discuss the various types of flooding, adaptation strategies, and the application of artificial intelligence. Global flooding has been extensively studied by the United Nations (UN), and this research is summarized, providing a rich trove of usable case studies. Advocacy briefs to legislators and courts are presented in useable detail. Land use is given special attention, along with how-to guidance for planners that can be implemented immediately.

Chapter Four

Storm Intensification

focuses on storm intensification. As global temperatures rise, so do the intensity, frequency, and unpredictability of storms. Climate change acts as an accelerant, warming oceans and saturating the atmosphere with moisture, which leads to more powerful hurricanes, cyclones, typhoons, and other extreme weather events. The winds at higher altitudes become more ragged and move in currently unpredictable ways. The amount of rain increases in amount and intensity, overwhelming drainage systems and drowning crops.

These intensified storms are no longer isolated anomalies; however, they are difficult to capture in current meteorological prediction models. They threaten lives, infrastructure, economies, and ecological systems worldwide. Creating climate adaptation systems for storm intensity is a high priority for climate adaptation generation. In this chapter, we analyze the case studies and the use of AI. Advocacy briefs to legislators and courts are presented in useable detail. Land use is given special attention, with guidance on how to approach it that can be implemented immediately.

Chapter Five

Drought

focuses on drought. There are many kinds of droughts. It is often described as a rolling natural disaster. Global warming dries out landscapes and introduces hotter temperatures for extended periods. The availability of water is key. The impact on the natural and built environment is severe. The World Meteorological Organization (2024) reports that over 3 billion people now live under water stress, and projections suggest that as many as 700 million individuals could be displaced by drought-related conditions by 2030. Case studies and the current application of AI programs to climate adaptation plans are described. Advocacy briefs to legislators and courts are presented in useable detail. Land use is given special attention, with guidance on how to approach it that can be implemented immediately.

Chapter Six

Wildfires

focuses on wildfires. Wildfires intensify due to prolonged droughts, extreme heat, and changing precipitation patterns—all of which are fueled by climate change. In urban areas, wildfires can release dangerous air contaminants as well as particulates. The Climate adaptation generation faces mounting pressure to develop proactive adaptation strategies that protect lives, ecosystems, and infrastructure. Wildfire adaptation necessitates a shift from reactive suppression to systemic, long-term resilience rooted in land-use planning, ecological restoration, and community risk reduction. Case studies and the current application of AI programs to wildfire detection, prevention, and preparation are described. Advocacy briefs to legislators and courts are presented in useable detail. Land use is given special attention, with guidance on how to approach it that can be implemented immediately.

Chapter Seven

Earthquakes

focuses on earthquakes. Earthquake adaptation involves preparing societies, infrastructure, and ecosystems to minimize damage, save lives, and facilitate rapid recovery in the event of seismic events. Unlike climate change-related adaptation, earthquake adaptation focuses on sudden-onset geophysical hazards. Earthquakes can last for days. Earthquakes can happen in unison with tsunamis and floods. An earthquake in an urban area can disrupt gas lines, leading to fires. Emergency responders must prepare for floods and fires in the aftermath of an earthquake. A rapidly developing set of AI programs is available for the climate adaptation generation discussed in this context. Case studies of their successful application are analyzed. As in the other chapters, advocacy briefs to legislators and courts are presented in useable detail. Land use is given special attention, with guidance on how to approach it that can be implemented immediately.

Chapter Six
Wildfires

focuses on wildfires. Wildfires intensify due to prolonged droughts, extreme heat, and changing precipitation patterns—all of which are fueled by climate change. In urban areas, wildfires can release dangerous air contaminants as well as particulates. The Climate adaptation generation faces mounting pressure to develop proactive adaptation strategies that protect lives, ecosystems, and infrastructure. Wildfire adaptation necessitates a shift from reactive suppression to systemic, long-term resilience rooted in land-use planning, ecological restoration, and community risk reduction. Case studies and the current application of AI programs to wildfire detection, prevention, and preparation are described. Advocacy briefs to legislators and courts are presented in useable detail. Land use is given special attention, with guidance on how to approach it that can be implemented immediately.

Chapter Seven
Earthquakes

focuses on earthquakes. Earthquake adaptation involves preparing societies, infrastructure, and ecosystems to minimize damage, save lives, and facilitate rapid recovery in the event of seismic events. Unlike climate change-related adaptation, earthquake adaptation focuses on sudden-onset geophysical hazards. Earthquakes can last for days. Earthquakes can happen in unison with tsunamis and floods. An earthquake in an urban area can disrupt gas lines, leading to fires. Emergency responders must prepare for floods and fires in the aftermath of an earthquake. A rapidly developing set of AI programs is available for the climate adaptation generation discussed in this context. Case studies of their successful application are analyzed. As in the other chapters, advocacy briefs to legislators and courts are presented in useable detail. Land use is given special attention, with guidance on how to approach it that can be implemented immediately.

Chapter Eight

Designing for Climate Adaptation: The Role of Place

Focuses on the role of Place and Place-based Planning. Place-based planning for adaptation to climate change is the emerging model for climate adaptation. It is the preferred model for climate adaptation generation because it brings results and gives direction for the future. Globally, it is not always affiliated with formal governments. Marginalized groups, such as landless populations and Indigenous Peoples, among others, have developed models of climate adaptation that have captured the surge of technological communication and AI in climate adaptation planning. This surge is a significant characteristic of the climate adaption generation. As in the other chapters, advocacy briefs to legislators and courts are presented in useable detail. Land use is given special attention, with guidance on how to approach it that can be implemented immediately.

Chapter Nine

Climate Education Supporting Learning, Adaptation, and Flourishing

examines the role of education. Issues of climate anxiety are critical because they can hinder the development of knowledgeable climate change leaders precisely when we need them most. My approach is multicultural, inclusionary, and interdisciplinary. That is the world the climate adaptation generation will face. It's possible to do this with the guidelines of most state and national curricular requirements. The role of technology, especially AI, is explored. Case studies, lesson plans, and directions for the future of climate education are laid out here. As in the other chapters, advocacy briefs to legislators and courts are presented in useable detail. Land use is given special attention, with guidance on how to approach it that can be implemented immediately.

Chapter Ten

Climate Adaptation and the Health of the Climate Adaptation Generation

Zeros in on the effects of climate change on public health. Climate anxiety is rampant and getting worse. With global warming, disease vectors become more widespread and resilient. In the context of rapid global urbanization and equally rapid increases in inorganic and organic waste streams, the quality of public health becomes degraded. The water quality plummets in many areas, if it is available at all. Waste discharges and illegal ocean dumping of all kinds of waste, including radioactive waste, are more the norm than the exception. The air quality from fires, growing industrial emissions, and a growing global middle class also degrade the air, becoming a ubiquitous public health threat. Public health impacts from land-destructive consumerism are reflected in the presence of plastic and forever chemicals that literally permeate our brains and our unborn children. Here, too, the application of AI, case studies, advocacy briefs, and planners' guides, ready for direct application, provide how-to direction designed for both present and future applications.

Chapter Eleven

Blueprint for Flourishing

focuses on the role of visioning and flourishing. Definitions, applications, and legal statuses are examined. Visioning is extremely important for climate change adaptation because flourishing means we don't just aim to return to the status quo but to go beyond that. As in the other chapters, advocacy briefs to legislators and courts are presented in useable detail. Land use is given special attention, with guidance on how to approach it that can be implemented immediately.

Chapter Twelve

Climate Change Impacts on the Natural Environment and Adaptation Measures

looks at the significant climate changes in the land, air, and water of the planet. Global warming affects ocean currents, jet streams, sea levels, and potentially earthquakes. Many of these impacts are well known.
Many are also unknown, such as the effect of the weight of seawater on the rise of ocean sea level rise, which affects the seismic stability of the sea floor. How ocean currents and changes in the jet steam affect storm intensification is unknown. The climate adaptation generation is especially interested in exactly when and where climate impacts will occur. That will require a working knowledge of the cumulative effects, whether they be synergistic, additive, negating, or nothing. And they will need more understanding of what is currently unknown, which is why education in the classroom is vitally important. I published a book on that topic this year. The animals and plants are indicators of climate change and react to it, sometimes by extinction and at other times by expanding their ranges. The natural system presents a myriad of issues for just one species to save. Us. Again, the role of AI is discussed with a focus on prevention, preparation, and recovery. My focus here is on bioregions and ecosystems. They are an accurate measure of global success at climate adaptation and test our species' resilience. A range of case studies is presented. Advocacy briefs addressing various issues are presented. Planners have a history of engaging with environmental issues, but here, that engagement is expanded and interacts with Planning guidance in chapters 2-7.

Chapter Thirteen

The Climate Adaptation Revolution and Evolution

raises the political and real landscape the climate generation will need to focus on. It describes the dominant types of governments and their political stances on climate change. The question is posed about whether climate adaptation is revolution or evolution. Here, there are essential, unavoidable, and uncomfortable conversations.

Capitalism is often blamed for the human contribution to climate change. But don't other forms of societal beliefs also provide fuel for human impacts on carbon dioxide emissions and global warming?

We will define and refine the concept of "capitalism" and compare and contrast it with Marxism. Another topic of discussion is the public's engagement with the power of the state or nation. Governments can be oppressive, and even when not, they are often not very relevant to the people.

As a legal concept in the United States, climate adaptation must be viewed as a legitimate and necessary public purpose. The Climate Generation answers this with a resounding "Of course." But the law doesn't, and the older generation is heavily vested in owning private property. Similarly, corporate and influential political lobbies are prevalent in the Finance, Insurance, and Real Estate (FIRE) sector. These groups are not immune to the impacts of climate change, and it's affecting the profitability of insurance companies and what they consider a natural disaster.

Chapter Fourteen

Climate Adaptation: Summary and a Blueprint

concludes with a summary of this book and a discussion of bioregional approaches to handle transboundary climate impacts as well as the digital divide. Each chapter concludes with a set of references. The book concludes with a glossary and index.

CHAPTER ONE
Adaptation Generation, Resilience Generation and Power

"The highest goal of education is not knowledge, but action—flourishing is a verb."
— Aristotle, paraphrased from *Nicomachean Ethics* (Original: *"Eudaimonia is the activity of the soul by virtue."*)

We are living in an era that history will remember not only for its environmental challenges but for how we choose to respond. This book was born from a profound conviction: that amid the turbulence of climate change, there exists a profound opportunity for humanity to adapt, heal, and flourish—together. As a veteran educator, highly published scholar, United Nations observer, and lifelong student of Nature, I have witnessed firsthand both the weight of this moment and the potential it carries.

The Adaptation Generation refers to today's youth, those born into a world already altered by rising seas, warming temperatures, and ecological uncertainty. As of 2024, over 1 billion children are estimated to be at "extremely high risk" from the impacts of climate change, including exposure to heatwaves, water scarcity, air pollution, and vector-borne diseases1. They are not passive victims of this inheritance. They are problem solvers, innovators, and bridge builders.

The political power of people under 25 has become a defining force in global climate, justice, and democratic movements. Although often excluded from formal decision-making due to age restrictions or entrenched hierarchies, youth today are leveraging new forms of influence that transcend traditional political pathways. Their political power is rooted in demographics, digital fluency, moral authority, and global interconnectedness—and it is growing rapidly.

A. Demographic Influence and Numbers of the Adaptation Generation

People under the age of 25 make up more than 40% of the global population. In many low- and middle-income countries, they are the majority.

This "youth bulge" gives them demographic leverage in elections, markets, and movements. In the U.S., Gen Z and young Millennials are now the most racially and ethnically diverse generations in history, and their voter turnout has surged. According to the Center for Information & Research on Civic Learning and Engagement (CIRCLE), about 27% of youth aged 18–29 voted in the 2022 midterm elections, a historically high turnout for a non-presidential year.

United States population pyramid

A wide base suggests population increase, narrow base indicates declining birth rates.

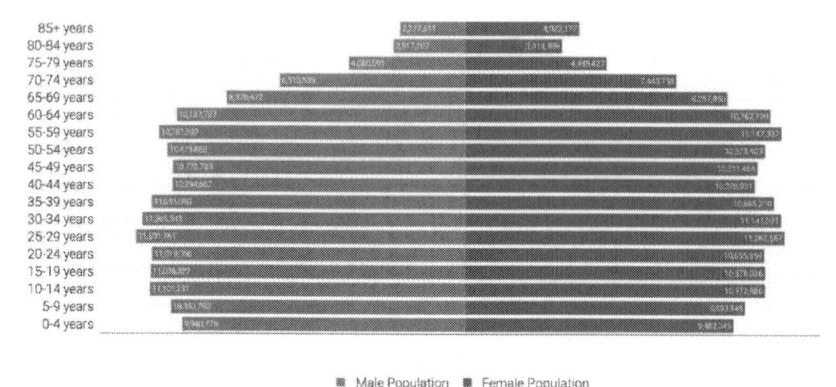

Male Population / Female Population chart by age group (85+ years down to 0-4 years)

Source: U.S. Census Bureau, American Community Survey (ACS) 2017-2021 5-Year Estimates

Neilsberg

Digital Power and Agenda Setting
Youth have unparalleled access to digital tools, allowing them to shape public discourse, organize mass mobilizations, and hold institutions accountable. Movements like Fridays for Future, March for Our Lives, Black Lives Matter, and #EndSARS in Nigeria were youth-led and used platforms like TikTok, Instagram, and Twitter/X to rapidly gain visibility and pressure leaders globally. The climate strike in September 2019 mobilized over 7.6 million people across 185 countries, making it one of the most significant youth-led political actions in history.

Moral Authority and Vision

Young people often frame their activism around intergenerational justice, demanding that governments and corporations act not only for today's interests but for their futures.

This gives them a powerful moral standing. Courts have increasingly recognized this. In 2023, Montana youth plaintiffs won a landmark climate case, Held v. Montana, in which the court ruled that the state had violated its constitutional right to a healthy environment. This case has become a model for youth climate litigation worldwide.

Civic Engagement and Innovation
While young people may distrust traditional institutions, they are innovating new forms of political participation: participatory budgeting, digital democracy platforms, mutual aid networks, and local organizing. Youth councils, climate assemblies, and school-based civic engagement programs are flourishing worldwide. In countries like Taiwan, digital participation tools for youth are integrated into national governance systems, setting a global precedent.

Voting Power and Electoral Influence
In pivotal elections, such as the 2020 and 2024 U.S. presidential elections, voters under the age of 25 have decisively influenced the outcomes. Exit polls from 2020 show that over 60% of voters aged 18–24 supported climate-forward candidates. In 2024, early polling indicated that youth were the deciding factor in swing states where climate, gun violence, and reproductive rights were top issues.

Global Youth Political Leadership
Around the world, young leaders are entering formal politics:

- Naisi Chen, elected to New Zealand's Parliament at age 26
- Greta Thunberg, while not an elected official, regularly addresses the United Nations and EU parliaments
- Bogolo Kenewendo, appointed Botswana's Minister of Investment at age 31, advocates for youth-driven green transitions
- Vanessa Nakate (Uganda), Licypriya Kangujam (India), and others are reshaping international climate diplomacy

Institutional Recognition and Support
Global bodies are beginning to recognize youth as a formal constituency:

- The United Nations Youth Envoy facilitates youth inclusion in global policy.
- The Youth4Climate platform, launched by UNDP and Italy, supports youth climate leadership.
- The Inter-Parliamentary Union (IPU) promotes quotas and training for young politicians, calling for parliaments to reflect the youth population they serve.

Adaptation Generation—today's youth under 25—by global income level:

- 65% live in middle-income countries
- 18% in low-income countries
- 17% in high-income countries

This highlights where most of the climate-affected youth are concentrated and where adaptation investments are most urgently needed.

B. Demographic Influence and Power of the Resilient Generation

The Resilience Generation includes those who came before—parents, educators, elders, and professionals—whose experience and accumulated wisdom are essential in shaping adaptive pathways forward. Globally, older adults are increasingly affected by climate-related disasters, with heatwaves alone contributing to 356,000 deaths of people over age 65 in 2019, up from 95,000 in 1990. Their insights are indispensable to preparing communities for what lies ahead.

The urgency of climate adaptation cannot be overstated. Since 2000, the number of weather-related disasters has increased by 83%, with over 7,000 major events recorded globally. In 2023, the United States experienced 28 separate billion-dollar weather and climate disasters, the highest ever recorded in a single year. Fires, floods, droughts, and displacements are no longer distant headlines—they are lived realities. But adaptation is not just about surviving. It's about thriving. It's about redesigning systems, rethinking education, and restoring community so that current and future generations can flourish even amid change.

The political power of people aged 65 and over is one of the most underappreciated yet profoundly influential forces in modern democratic societies, especially in the United States and other aging nations. This age group wields considerable political, economic, and civic influence, shaped by high voter turnout, wealth concentration, institutional memory, and a strong presence in local governance and advocacy.

Voting Power and Turnout
Older adults consistently have the highest voter turnout rates. In the United States, more than 70% of people aged 65 and older vote in national elections, which is significantly higher than any other age cohort. In contrast, fewer than 50% of voters under 30 typically cast ballots. This consistent engagement gives older voters disproportionate influence over election outcomes, especially in primary elections and ballot initiatives.

Demographic Growth
Globally and nationally, the population over 65 is growing. In the U.S., the number of seniors is projected to exceed 80 million by 2040—making them nearly one-fifth of the population. Countries such as Japan, Italy, and Germany already have aging populations that influence national policy agendas.

Wealth and Economic Influence

People over 65 control a large share of financial capital and real estate. In the U.S., they hold over 30% of all household wealth, and their economic power extends to campaign contributions, lobbying, and funding of political causes. Organizations like AARP and senior advocacy coalitions have leveraged this to influence policies on Social Security, healthcare, housing, and retirement protections.

Institutional Memory and Governance

Older adults often lead local civic institutions, school boards, planning commissions, neighborhood associations, and faith-based organizations. Their long-term historical memory helps preserve community identity, resist harmful development trends, and guide intergenerational planning. Elders are frequently the stewards of community resilience, disaster recovery efforts, and cultural preservation.

Policy Agenda Influence

Senior voters tend to prioritize stable healthcare, affordable housing, public safety, and concerns about the cost of living. However, there is growing support among older adults for climate action, generational equity, and sustainability, especially among Baby Boomers who participated in early environmental movements. The AARP and other organizations have supported energy efficiency, aging-in-place retrofits, and resilience infrastructure.

Climate and Intergenerational Power

Increasingly, older generations are becoming key allies in climate justice, adaptation planning, and youth empowerment. This cohort is essential for bridging historical knowledge with future visioning. Programs that involve elders in mentoring, climate storytelling, and intergenerational resilience projects show that seniors are not barriers to progress; they are catalysts.

Political Representation

A large portion of elected officials are over 65, from city council members to U.S. senators. While this brings experience, it also raises questions about how well institutions represent the voices of younger generations. However, the trend also offers opportunities to use elder leadership for transformative policy, particularly when guided by ethical frameworks, public service values, and a sense of legacy.

Risks of Generational Misalignment
There is potential for political polarization if younger generations feel ignored or if older generations resist change. However, where intergenerational collaboration is encouraged—through participatory planning, climate resilience summits, or shared leadership models—the political power of seniors can be stabilized and unified, thereby strengthening movements.

The 65+ population are a cornerstone of political power, both through institutional roles and grassroots influence. Rather than viewing aging populations as passive or conservative by default, recognizing their potential as mentors, advocates, and protectors of future generations reframes elders as essential leaders in an era of ecological, social, and democratic transition.

C. Environmental Context: Wastes and Disasters

In addition to climate-related catastrophes, the world faces a mounting crisis of accumulated waste. As of 2023, the planet has generated over 2.5 billion metric tons of solid waste annually, with projections exceeding 3.4 billion metric tons by 2050. Cumulatively, it is estimated that over 60 billion metric tons of waste, including plastics, industrial chemicals, municipal garbage, and toxic byproducts—have been dumped or inadequately managed across land, oceans, and waterways. This includes over 400 million tons of hazardous waste produced annually, much of which is concentrated in vulnerable communities and ecosystems. Electronic waste is a fast-growing contributor, with 50 million tons generated annually, yet only 20% is formally recycled.

These toxic legacies endanger biodiversity, contaminate drinking water, and contribute to long-term public health crises, particularly in low-income, Indigenous, and marginalized areas. Addressing this burden is essential not only for environmental survival but also for moral justice. Accurate adaptation must confront both atmospheric and material pollution to ensure a flourishing future.

Projected Global Solid Waste Generation (2020–2050) This visualization highlights the escalating volume of solid waste globally, projected to surpass 3.4 billion metric tons by 2050.

Billion-Dollar Weather and Climate Disasters in the U.S. (2000–2023)
This graph illustrates the increasing number of costly climate-related disasters, which peaked at 28 in 2023.

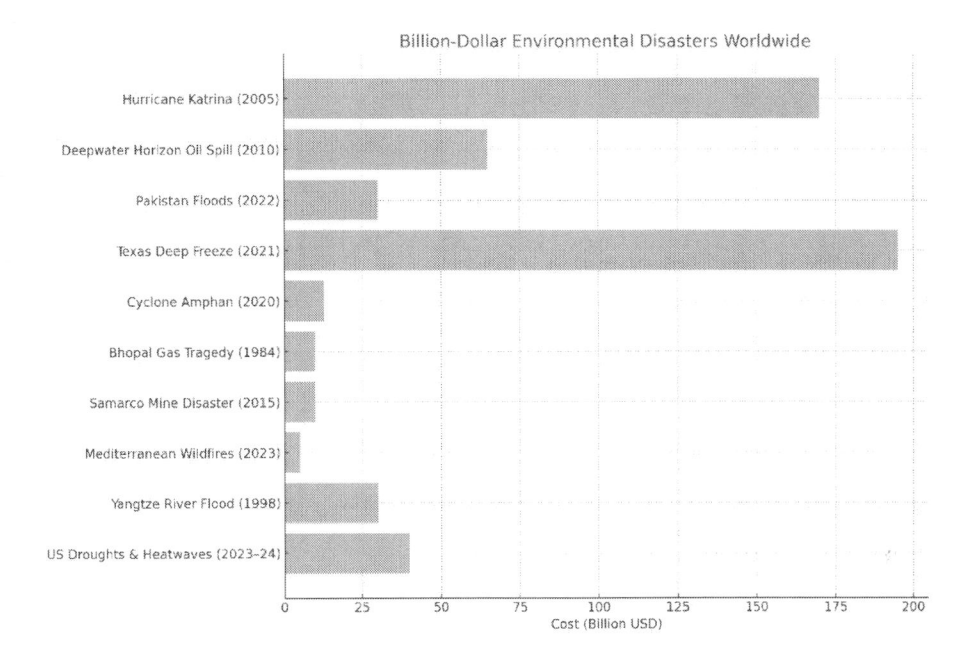

Billion-Dollar Environmental Disasters Worldwide

Disaster	Cost (Billion USD)
Hurricane Katrina (2005)	~170
Deepwater Horizon Oil Spill (2010)	~65
Pakistan Floods (2022)	~30
Texas Deep Freeze (2021)	~195
Cyclone Amphan (2020)	~15
Bhopal Gas Tragedy (1984)	~10
Samarco Mine Disaster (2015)	~12
Mediterranean Wildfires (2023)	~8
Yangtze River Flood (1998)	~30
US Droughts & Heatwaves (2023-24)	~40

Exposure to Hazardous Waste by Population Group. It indicates that 60% of hazardous waste exposure affects vulnerable and marginalized communities.

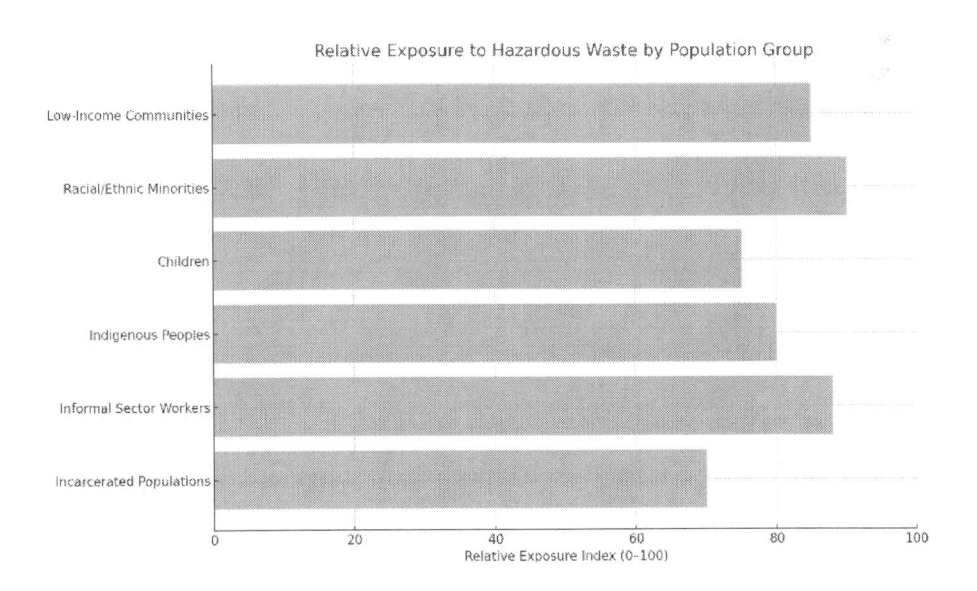

Relative Exposure to Hazardous Waste by Population Group

Population Group	Relative Exposure Index (0-100)
Low-Income Communities	~85
Racial/Ethnic Minorities	~90
Children	~75
Indigenous Peoples	~80
Informal Sector Workers	~88
Incarcerated Populations	~70

D. Flourishing is the Goal for Everyone

Flourishing, as described by the Harvard Human Flourishing Program, encompasses more than physical wellbeing or economic stability. It encompasses six domains: happiness and life satisfaction, mental and physical health, meaning and purpose, character and virtue, close social relationships, and financial and material stability. In the context of climate change, flourishing means equipping individuals and communities not only to withstand disruption but to live lives of purpose, connection, and joy despite uncertainty. It means cultivating environments—ecological, emotional, and educational—where young people feel empowered and communities feel supported.

Flourishing also requires emotional resilience: the ability to face ecological grief, climate anxiety, and disruption with clarity and compassion. It involves educational systems that nurture critical thinking and creativity, leadership that is empathetic and inclusive, and community design that is regenerative and equitable. These conditions are not accidental—they must be cultivated intentionally through policies, practices, and collective care.

The Harvard Flourishing Study, developed by Dr. Tyler J. VanderWeele and his team at the Human Flourishing Program at Harvard University, offers a comprehensive framework for measuring and promoting well-being. It defines flourishing through six interconnected domains: happiness and life satisfaction, mental and physical health, meaning and purpose, character and virtue, social relationships, and financial and material stability. When extended to global populations, the framework becomes not only a measurement tool but a transformative approach to shaping public health, education, economic development, climate adaptation, and spiritual life across diverse cultures.

Cross-culturally, the flourishing framework reveals both the universality and cultural specificity of well-being. In practice, the domains have been adapted across linguistic and regional contexts, from Asia to Africa and Latin America. This has shown that while the core dimensions of human flourishing are shared, their meanings vary. For instance, a sense of meaning and purpose may arise from religious devotion in West Africa. At the same time, it may stem from professional autonomy in Northern Europe.

The framework allows for these differences without imposing a Western-centric model. Bhutan's Gross National Happiness index aligns with the flourishing domains, providing a notable example of a policy built around holistic well-being. This signals the possibility for flourishing-based metrics to serve as decolonized alternatives to GDP or even the Human Development Index.

In global public health, the flourishing model reshapes our understanding of healing and thriving. By emphasizing the integration of mind, body, purpose, and social connection, it supports more person-centered and holistic care. The World Health Organization has incorporated flourishing concepts into trauma recovery programs in conflict and disaster-affected regions such as Syria and Haiti. In Sub-Saharan Africa, flourishing domains are being measured alongside HIV/AIDS metrics to assess not only biological health outcomes but also life satisfaction, hope, and relational support systems. This more comprehensive assessment provides a more complete picture of human recovery and resilience. In the context of long COVID, various measures have been employed to assess whether individuals have truly returned to a state of thriving—not only in clinical terms but also in terms of regained meaning and relational well-being.

Globally, education systems have begun to integrate flourishing principles into youth development frameworks. UNESCO and other international networks have adopted the flourishing model to assess student well-being, especially in high-risk or conflict-affected zones. In Colombia and Lebanon, for example, flourishing surveys have been used to understand resilience and character development in displaced and refugee youth. This shift reflects an educational philosophy that values more than academic achievement; it promotes social and emotional learning, ethical formation, and relational well-being. In refugee education contexts, domains like meaning, virtue, and community often predict success more accurately than grades or test scores.

The flourishing framework is also instrumental in reimagining aging in a global context. As countries such as Japan, South Korea, and Italy face rapidly aging populations, flourishing provides a broader set of indicators to measure the well-being of older adults.

In Japan, for example, flourishing indicators are used to evaluate the success of inclusion programs that connect elderly individuals to social networks, purpose-driven activities, and intergenerational projects. In Uganda and the Netherlands, intergenerational housing initiatives use flourishing indicators to show that relationships and shared meaning benefit both elders and youth. The framework not only evaluates life in old age but supports policies that increase dignity, agency, and contribution across generations.

Dimensions of Human Flourishing

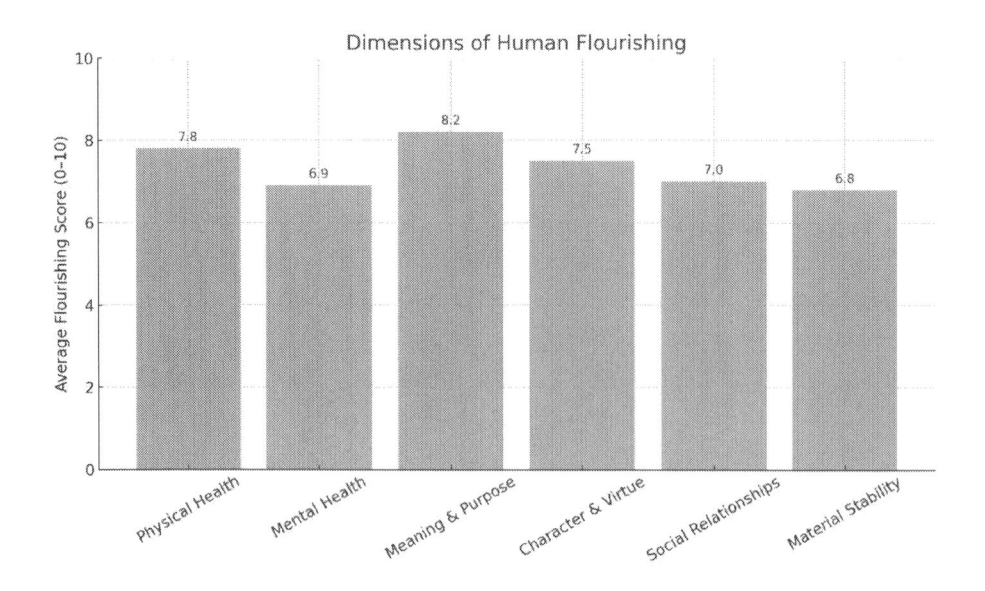

Dimension	Score
Physical Health	7.8
Mental Health	6.9
Meaning & Purpose	8.2
Character & Virtue	7.5
Social Relationships	7.0
Material Stability	6.8

As climate change intensifies, the concept of flourishing is being applied to adaptation and resilience strategies. Governments and NGOs in climate-vulnerable areas are using flourishing indicators to measure the human impacts of disasters and displacement. Following the 2022 floods in Pakistan and the 2023 heatwaves in India, assessments were conducted to gauge not only physical survival but also psychological recovery, communal cohesion, and the restoration of purpose. This application of flourishing reorients climate adaptation toward long-term, value-driven recovery rather than simply rebuilding infrastructure. It also supports the development of just transitions, ensuring that communities not only survive environmental change but can reconstruct their lives with dignity, meaning, and connectedness.

In the economic sphere, flourishing is reshaping global development models by questioning GDP as the dominant measure of success.

The OECD's Better Life Index and the UN Sustainable Development Goals are now incorporating flourishing indicators into their assessments. In regions such as South Asia and East Africa, microloan programs and women's cooperatives are utilizing flourishing surveys to measure empowerment, meaning, and relational well-being alongside income gains. This creates a more ethical and human-centered approach to economic development, in which financial stability is not the endpoint but one of many conditions required for thriving.

Religious and ethical traditions around the world have also begun to engage in the flourishing model as a spiritually resonant framework. Faith-based organizations across Latin America, Africa, and Southeast Asia have incorporated flourishing surveys into pastoral care, community development, and moral education. In Muslim, Christian, and Buddhist communities, the domains of meaning, virtue, and social connection find natural resonance with spiritual teachings. In Thailand, for instance, Buddhist monastic schools use flourishing principles in youth training. In contrast, churches in Kenya and Brazil use the framework to design community support programs. The flourishing model is values-neutral yet compatible with diverse worldviews, allowing spiritual leaders to promote both well-being and ethical development.

The data infrastructure emerging from the flourishing model is also shaping global policy. The Global Flourishing Study, a collaboration among Harvard, Baylor University, and Gallup, is currently tracking over 240,000 people in 22 countries through 2030. This dataset is being used to inform global well-being indices, national policy, and development planning. Nations such as Finland, Costa Rica, and Rwanda have begun incorporating flourishing-based indicators into public reporting and strategic governance, aligning policies not just with economic outcomes but with holistic measures of societal health.

In conclusion, the Harvard Flourishing Study is not just a research tool but a transformative framework for understanding human well-being across global populations. Its domains provide a shared language for health, purpose, ethics, education, and governance—while allowing cultural flexibility and spiritual depth. It supports a shift from survival-based metrics to dignity-centered development, offering a moral compass for the climate era and a foundation for just, inclusive, and sustainable futures.

E. The Role of Artificial Intelligence in Human Flourishing in a Changing Climate

Artificial Intelligence (AI) has the potential not only to mitigate climate disasters but to actively contribute to human flourishing—the full realization of individual and collective wellbeing, purpose, and resilience—especially in the context of accelerating environmental change. Rooted in ethical innovation, participatory governance, and climate justice, AI can support societies in adapting to climate risks while enhancing human potential, dignity, and agency.

1. Supporting Physical and Mental Health

AI-powered systems are increasingly used to monitor and address the public health impacts of climate change, including heatwaves, air pollution, waterborne diseases, and displacement stress. Predictive analytics help identify emerging health threats in vulnerable communities, while AI-driven telehealth platforms increase access to care in remote or flood-prone regions.

By integrating mental health screening tools with climate impact data, AI can also identify psychosocial risks, such as climate anxiety, post-disaster trauma, and eco-grief, and provide early interventions. This supports emotional resilience, a critical component of flourishing as defined by the Harvard Flourishing Study.

2. Enhancing Education and Climate Literacy

AI tools can personalize climate education, making learning adaptive, multilingual, and culturally relevant. For example, AI tutors and interactive simulations can guide students through complex climate systems, resilience planning, and ethical decision-making. These educational experiences foster a sense of agency and ecological identity, both of which are essential to personal development and community flourishing.

In underserved areas, AI expands access to climate education by overcoming teacher shortages, language barriers, and curriculum gaps—empowering the next generation to engage with hope, science, and creativity.

3. Empowering Adaptive Governance and Participation

AI can democratize climate governance by analyzing large datasets (e.g., flood risks, emissions, housing inequity) and visualizing solutions for citizens and planners. Tools like participatory scenario modeling and AI-enabled visioning exercises allow communities to co-create plans rather than being passive recipients of top-down decisions.

Such inclusion is foundational to flourishing, ensuring people feel heard, empowered, and capable of shaping their future—even amid environmental challenges.

4. Preserving Culture and Memory

AI is also being used to archive Indigenous ecological knowledge, oral histories, and place-based adaptation practices—many of which are at risk due to rising seas, wildfires, and migration. By digitizing these cultural assets, AI helps sustain identity and intergenerational continuity, key ingredients of human flourishing.

For displaced or frontline communities, AI can support cultural resilience by mapping lost homelands, tracing ancestry, or recreating ceremonial spaces virtually. These technologies reinforce a sense of belonging and purpose amid disruption.

5. Designing for Flourishing Cities

Smart cities increasingly use AI to optimize green space access, water use, food systems, transportation, and housing, which directly affect quality of life. AI can identify inequities in access to Nature, cooling, or healthy food and help planners redesign urban ecosystems that support health, mobility, and dignity.

AI-powered environmental sensors enable real-time monitoring of air quality, temperature, flood threats, and noise, allowing cities to protect and enhance the lived experience for all residents.

One of the most expansive initiatives is the Harvard Human Flourishing Program's AI & Wellbeing collaboration, which explores how machine learning can be used to track the well-being of populations over time.

Through data from health records, surveys, and environmental sensors, the program models are indicators of flourishing—including purpose, health, relationships, and resilience—and help governments and health systems intervene before well-being deteriorates. These tools support proactive, rather than reactive, public health.

In climate-impacted areas, UNICEF's AI for Resilient Futures program uses predictive analytics to anticipate risks to children's education, mental health, and housing. AI systems forecast how floods, heatwaves, and food insecurity might disrupt schooling and wellbeing, allowing humanitarian agencies to deliver early interventions that sustain opportunity and hope—critical elements of flourishing for youth.

Wadhwani AI, based in India, utilizes machine learning to enhance rural health systems by detecting early signs of anemia, malnutrition, and disease outbreaks in underserved communities.

These tools empower local health workers with real-time diagnostic support, contributing to a healthier, more equitable foundation for wellbeing.

In Kenya and Bangladesh, the AI for Dignified Work and Sustainable Livelihoods platform, supported by the UNDP, assists smallholder farmers in adapting to climate shocks by analyzing weather trends, crop yields, and soil data. These insights are translated into accessible recommendations—via text messages or radio—that help preserve income, food security, and a sense of control over one's future.

For emotional and psychological well-being, Woebot Health AI provides mental health support through conversational AI trained in cognitive-behavioral techniques. While initially designed for anxiety and depression, it has been increasingly used to support climate-related stress and displacement trauma. The system is integrated into community programs in Australia and the U.S., providing accessible, stigma-free support.

UNESCO's Indigenous Knowledge AI Lab is preserving endangered languages, traditional ecological practices, and oral histories through natural language processing and AI-powered archiving. This cultural preservation fosters identity, continuity, and intergenerational belonging—critical dimensions of human flourishing in communities facing climate displacement.

In cities like Helsinki and Amsterdam, AI for Democratic Participation platforms are helping residents co-design policies, vote on local budgets, and shape urban resilience strategies. These systems use machine learning to synthesize citizen input, generate equitable proposals, and reduce planning bias. By empowering participation, these platforms expand civic agency and shared purpose.

Meanwhile, AI for Nature Connectedness, a growing interdisciplinary field, utilizes algorithms to guide people toward more sustainable behaviors and increased interaction with the natural world, recommending walking routes through biodiverse areas or matching urban residents with stewardship projects. These initiatives contribute to emotional wellbeing and a sense of planetary belonging.

In summary, these programs reflect a growing awareness that AI, when guided by human values and principles of justice, can serve not only to mitigate harm but also to cultivate *well-being, meaning, belonging, and purpose—the essential components of flourishing.*

F. Conclusion

Intergenerational responsibility lies at the heart of this transformation. We must honor the knowledge of resilient older adults while empowering the energy of climate-adaptive youth. We must listen, learn, and work together to be led. This is not a book of despair. It is a book of possibility. It is for the teacher to inspire, the student seeking clarity, the parent yearning for practical tools, the planner looking for guidance, the advocate for climate adaptation, and the leader ready to act.

Let us reimagine the climate future as one of collective strength, innovation, and compassion. Let us raise generations who not only endure change but help shape a world that is more just, sustainable, and alive with hope—and who flourish, even in the face of adversity.

Advocacy Brief

Advancing Climate Adaptation through Human Flourishing

Overview:
Climate adaptation efforts must go beyond infrastructure and emissions to prioritize human well-being and long-term flourishing. As climate disruptions escalate—from extreme heat to displacement—adaptation strategies that center dignity, equity, and opportunity are essential. A flourishing-centered approach measures success not just by survival, but by thriving communities.

Policy Challenge:
Traditional adaptation metrics focus narrowly on risk reduction and infrastructure resilience. However, they often neglect mental health, cultural vitality, democratic participation, and intergenerational equity—key components of human flourishing.

Strategic Opportunity:
Integrating flourishing frameworks—such as the Harvard Human Flourishing Program and the OECD Well-Being Indicators—into climate adaptation planning enables policies that address social determinants of resilience, including housing stability, access to nature, education, and community cohesion.

Action Steps for Policymakers and Planners:
1. *Incorporate validated flourishing indicators into climate adaptation assessments.*
2. *Co-design policies with marginalized groups to ensure culturally relevant, community-led solutions.*
3. *Invest in adaptation projects that support ecological health and human potential, such as green schools, local food systems, and participatory planning.*

Conclusion:
Flourishing is not a luxury—it is the foundation of sustainable adaptation. As we reimagine a climate-resilient future, policies must empower communities not only to endure, but to grow with purpose, connection, and hope.

Planners Toolkit

Climate Adaptation for Human Flourishing

1. *Flourishing Indicators Integration*
   Use metrics from the Harvard Flourishing Measure and OECD Well-Being Framework to assess adaptation success in areas like purpose, health, relationships, and environmental quality.

2. *Visioning Workshops*
   Facilitate inclusive, community-led scenario planning that centers local values, cultural knowledge, and long-term aspirations.

3. *Equity Mapping*
   Overlay climate risk data with social vulnerability indicators (income, housing, race, health) to prioritize adaptation investments that close equity gaps.

4. *Nature-Based and Social Infrastructure*
   Design green spaces, cooling centers, and resilient schools that foster both ecological function and community well-being.

5. *Multi-Sector Collaboration*
   Coordinate with public health, education, housing, and transportation agencies to build adaptation strategies that support whole-of-life flourishing.

6. *Feedback Loops*
   Implement participatory evaluation tools (e.g., community scorecards, storytelling, youth councils) to adapt plans over time based on lived experience.

Goal

Build adaptive communities where people are not just safe, but empowered to lead meaningful, connected, and purpose-driven lives in a changing climate.

RESOURCES

1. UNICEF. (2021). The Climate Crisis Is a Child Rights Crisis: Introducing the Children's Climate Risk Index. https://www.unicef.org/reports/climate-crisis-child-rights-crisis
2. Romanello, M. et al. (2021). The 2021 report of the Lancet Countdown on health and climate change: code red for a healthy future. The Lancet, 398(10311), 1619–1662. https://doi.org/10.1016/S0140-6736(21)01787-6
3. Centre for Research on the Epidemiology of Disasters (CRED). (2020). The Human Cost of Disasters: An Overview of the Last 20 Years (2000–2019). https://www.undrr.org/publication/human-cost-disasters-overview-last-20-years-2000-2019
4. National Oceanic and Atmospheric Administration (NOAA). (2024). Billion-Dollar Weather and Climate Disasters. https://www.ncei.noaa.gov/access/billions
5. VanderWeele, T. J. (2017). On the promotion of human flourishing. Proceedings of the National Academy of Sciences, 114(31), 8148–8156. https://doi.org/10.1073/pnas.1702996114
6. CIRCLE. (2023). Youth Voting in 2022 Midterms. Tufts University.
7. United Nations Population Fund. (2024). State of World Population.
8. Held v. Montana, 2023 MT 204 (Montana First Judicial District Court).
9. Pew Research Center. (2024). Gen Z's Political Engagement Patterns.
10. IPU. (2025). Youth Participation in National Parliaments.
11. UNDP & Italy. (2024). Youth4Climate: Call for Solutions Report.
12. Harvard Kennedy School. (2025). Digital Participation and Youth Democracy.

Chapter Two
Emergency Responders and Climate Change

"Emergency responders are no longer just the last line of defense—they are the first architects of resilience. In the era of climate change, every call answered is not just a response to crisis, but a commitment to adaptation."
Robert W. Collin

A. Overview

As climate change accelerates, the frequency and severity of natural disasters, including wildfires, hurricanes, floods, and heatwaves, emergency response systems are under growing pressure. Artificial Intelligence (AI) provides crucial tools to enhance the speed, precision, and coordination of disaster preparedness and response. From early warning systems to real-time rescue coordination, AI is reshaping how responders prepare for and react to climate-related crises.

1. The Role of Emergency Responders in Climate Adaptation

Emergency responders, firefighters, paramedics, disaster relief personnel, search and rescue teams, and emergency management agencies—are on the front lines of climate adaptation. As climate change intensifies, the frequency and severity of disasters, such as wildfires, floods, heatwaves, and hurricanes, increases. Consequently, the role of these responders has evolved beyond immediate crisis management into proactive climate resilience planning and implementation.

1. Frontline Observation and Data Collection

Emergency responders are often the first to observe emerging patterns of risk in real-time. Their field reports, sensor data, and incident logs provide invaluable localized insights into how climate change is reshaping hazard profiles. In many regions, this situational intelligence is increasingly used to inform municipal adaptation strategies, land-use planning, and infrastructure upgrades.
Example: In California, wildfire responders are equipping drones and mobile sensors to collect data on fire spread patterns, which is then shared with urban planners and environmental scientists for adaptation modeling.

2. Early Warning and Public Communication

Responders play a crucial role in disseminating early warning information during climate-related events. They often collaborate with meteorological agencies and AI-driven forecasting systems to ensure that alerts reach vulnerable populations—particularly in low-income or marginalized communities with limited access to digital infrastructure.
Example: During Hurricane Ian (2022), Florida emergency teams used SMS alerts, multilingual audio broadcasts, and door-to-door canvassing to reach elderly and disabled residents in evacuation zones.

3. Community Risk Assessment and Climate Literacy

Emergency responders increasingly participate in public education campaigns that promote climate risk awareness. Through drills, school programs, and neighborhood workshops, they help communities understand their vulnerability and prepare adaptation strategies, such as creating evacuation routes, defensible space, and preparing go-bags.

Example: Japan's Disaster Preparedness Day involves first responders staging simulations and educating families about earthquake and tsunami safety, which also promotes long-term adaptation mindsets.

4. Integrated Infrastructure Planning and Climate Resilience

In collaboration with urban planners and climate scientists, emergency responders now contribute to the design of infrastructure that can withstand future climatic extremes. Their feedback is critical for developing cooling centers, firebreaks, stormwater systems, resilient shelters, and elevated roadways.

Example: Following the 2021 floods in British Columbia, Canadian emergency personnel were integrated into recovery planning teams to redesign rural access roads, taking into account increased landslide and flood risk due to atmospheric river events.

5. Mental Health Support and Trauma Recovery

Climate disasters often have long-term mental health consequences. Emergency responders are increasingly being trained in trauma-informed care, post-disaster counseling, and climate grief support.

This human-centered adaptation role addresses the psychological impacts of climate instability, particularly on children and older people.

Example: In Australia, paramedics and firefighters participated in mental health outreach following the Black Summer bushfires (2019–2020), working with psychologists to stabilize traumatized communities.

6. AI and Predictive Modeling Integration

As climate risks become increasingly complex, responders are utilizing AI tools for real-time disaster mapping, resource allocation, and vulnerability assessment. This technology helps anticipate disaster intensity, optimize emergency logistics, and simulate future scenarios to improve response capacity.

Example: FEMA's AI-enhanced flood modeling systems are utilized by emergency teams to identify pre-deployment zones, thereby reducing response times and ensuring that critical areas are prioritized during storms.

7. Equity-Centered Adaptation

Responders are being trained to recognize environmental justice dimensions—ensuring that vulnerable groups, such as unhoused populations, Indigenous communities, or migrant workers, receive equitable assistance before, during, and after climate-related emergencies. Climate adaptation must be inclusive to succeed.
Example: In Phoenix, Arizona, first responders partner with social workers to provide hydration stations, mobile cooling units, and wellness checks during extreme heat events—especially targeting elderly and low-income residents.

Emergency responders are no longer just reactive actors; they are key agents in proactive climate adaptation. Their integration into planning, education, equity, and technology deployment ensures that climate resilience is not just a theoretical framework but a lived reality for communities under threat. Empowering and equipping responders with climate-specific tools, training, and cross-sector partnerships is essential for adaptive capacity in the face of escalating environmental risks.

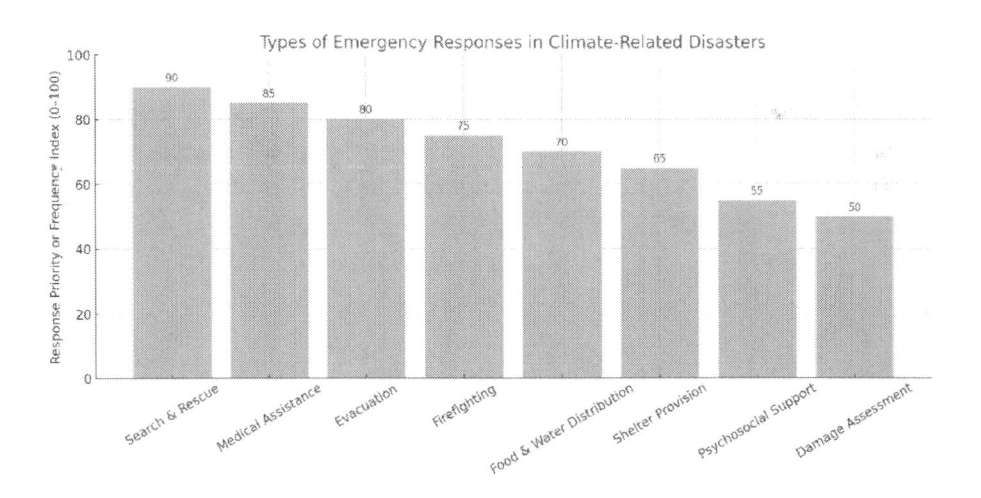

Types of Emergency Responses in Climate-Related Disasters

Response Type	Response Priority or Frequency Index (0-100)
Search & Rescue	90
Medical Assistance	85
Evacuation	80
Firefighting	75
Food & Water Distribution	70
Shelter Provision	65
Psychosocial Support	55
Damage Assessment	50

B. Applications of AI in Emergency Responses to Climate Events

Early warning and forecasting systems are undergoing a revolutionary transformation with the integration of artificial intelligence (AI), which enhances the precision, speed, and adaptability of climate disaster responses. One of the most critical advancements is in AI-enhanced weather modeling.

These models assimilate vast quantities of real-time climate, geospatial, and satellite data to enhance the prediction of extreme events, including hurricanes, floods, and wildfires. By employing machine learning algorithms that detect anomalies and emerging patterns in seismic, oceanic, and atmospheric datasets, these systems can provide crucial early warnings—ranging from minutes to hours—before disasters strike. For example, IBM's Watson AI system is being leveraged to analyze the formation of tropical storms and cyclones, offering more accurate predictions about landfall zones in vulnerable regions. This enables faster mobilization of resources and more targeted evacuations.

AI is also reshaping how disaster risk is mapped and how vulnerability assessments are conducted. By analyzing satellite imagery, demographic profiles, land use patterns, and topographic data, AI systems can identify high-risk zones with greater granularity than traditional tools. These predictive models are instrumental for governments and urban planners seeking to prioritize infrastructure investments, emergency planning, and the development of resilience hubs. A notable initiative is Google's AI for Social Good, which maps flood-prone areas in India and Bangladesh, offering hyper-local alert systems that reach communities before rivers breach their banks.

In real-time disaster scenarios, AI is proving indispensable in crisis management. During emergencies, AI-powered platforms analyze incoming data from emergency calls, social media activity, and sensor networks to pinpoint the location of people in distress. Natural Language Processing (NLP) algorithms allow these systems to rapidly interpret and prioritize messages from platforms like Twitter or SMS, distinguishing genuine emergencies from misinformation. Projects such as CrisisNET and Ushahidi exemplify how crowdsourced information can be verified and deployed effectively for first responders using AI tools, improving the accuracy and timeliness of interventions.

Drones and robotic systems are another frontier where AI is proving transformative. Equipped with computer vision and advanced navigation algorithms, autonomous drones can survey damaged terrain, assess compromised infrastructure, and deliver essential supplies to areas cut off from aid. Ground-based robots, guided by AI, support search-and-rescue teams in hazardous environments, reducing human exposure to risk. During the California wildfires, AI-directed drones were used extensively to map active fire lines and assist firefighting crews in deploying resources more efficiently and safely.

AI is also playing a vital role in predicting and managing climate-driven disease outbreaks in the aftermath of disasters. Changes in rainfall, temperature, and sanitation conditions can trigger outbreaks of diseases like cholera, dengue, or malaria. AI models can analyze these factors in real time to forecast where and when outbreaks are likely to occur. This enables health agencies to deploy medical resources preemptively and design targeted public health campaigns, especially in post-disaster settings where communities are most vulnerable.

In terms of preparedness, AI-enhanced training and simulation environments are elevating the capacity of emergency responders. Simulations powered by AI allow teams to rehearse complex disaster scenarios, such as flash floods in urban environments or cascading wildfire events. These virtual environments adapt dynamically to user inputs, mimicking the unpredictability of real-world disasters. Such training not only builds individual skills but also strengthens coordination across response teams, significantly increasing readiness for rare and high-impact events.

Real-world case studies offer further evidence of AI's growing importance. In the United States, the Federal Emergency Management Agency (FEMA) has adopted AI tools to optimize resource allocation during hurricane seasons, predict the demand for hospitals and shelters, and manage supply logistics. Kenya provides a compelling example of AI being utilized to combat drought; here, AI models, paired with satellite data, predict pastoralist migration patterns and assess livestock health, thereby informing food and water distribution strategies in arid regions. In Canada, the FireGuard AI system has emerged as a critical tool in wildfire management, using inputs such as wind speed, humidity, and terrain data to predict fire behavior and inform timely evacuation decisions.

Together, these developments demonstrate that AI is not merely a technological enhancement but a vital tool in the global response to climate-related disasters. It is reshaping how we anticipate, manage, and recover from crises—bringing new hope to communities on the frontlines of climate change.

C. Future Potential and Global Recommendations

The integration of AI into climate adaptation and disaster response is not only a technological opportunity but a planetary necessity. To maximize its potential, AI-powered platforms must be embedded within both national and local disaster management systems.

These systems—from early warning networks to emergency coordination centers—should be upgraded with AI tools capable of processing vast datasets, forecasting extreme events, and supporting real-time decision-making. Countries that invest in such AI-climate platforms will be better equipped to respond swiftly to wildfires, hurricanes, floods, and other intensifying climate-related hazards.

Global equity must remain a guiding principle in this transformation. Open-source AI tools should be promoted and funded to ensure that nations in the Global South can access, adapt, and scale technological solutions without prohibitive costs. Without this, AI risks deepening existing disparities in climate resilience, leaving the most vulnerable regions further exposed.

In parallel, there must be substantial investment in training for human-AI collaboration. Emergency responders, public health officials, urban planners, and civil defense teams must understand not only how to utilize AI tools but also how to interpret their output critically, act on probabilistic models, and communicate decisions ethically during high-pressure crises. Human judgment remains essential, and AI must enhance—not replace—local expertise and cultural knowledge.

Finally, transparent standards must be established for the deployment of AI in disaster governance. This includes developing international norms for data privacy, bias mitigation, system explainability, and public accountability. People have a right to understand how AI is influencing decisions that affect their safety, mobility, and survival. Transparent and participatory frameworks are essential for building public trust and ensuring that AI becomes a tool of empowerment rather than exclusion.

Together, these steps form the foundation of a future where AI serves as a strategic partner in confronting climate emergencies—one that is inclusive, responsible, and globally aligned.

As climate-related disasters intensify in frequency, scope, and unpredictability, artificial intelligence is becoming a central pillar in transforming emergency response systems. Looking ahead, AI will increasingly serve not only as a reactive tool during crises but as a proactive, anticipatory system capable of guiding policy, prevention, and preparedness at multiple scales (UNDRR, 2025).

One of the most significant trends is the emergence of hyper-predictive systems powered by climate-aware AI. These systems integrate data from Earth system models, near-real-time satellite streams, local climate variables such as soil moisture and heat indices, and socioeconomic vulnerability maps. This integration enables predictive intelligence that can forecast compound climate risks—such as the convergence of drought, heat, and social vulnerability—days or even weeks in advance (Cruz & Kumar, 2024). These models also support real-time digital twin environments, allowing emergency responders to visualize disaster progression by the hour and dynamically adjust evacuation plans and resource distribution (FEMA AI Futures Lab, 2025).

AI is also shifting toward decentralized, edge-based computing. This evolution empowers local responders and communities with on-device intelligence that does not rely on cloud infrastructure—a critical innovation for areas with low connectivity or those compromised by disasters. Edge AI devices embedded in mobile phones, solar radios, or local weather stations enable real-time decision-making, even in offline settings (World Bank, 2024). This trend will be particularly valuable in island nations and rural regions where centralized systems are vulnerable during extreme weather events.

Another transformative development is the convergence of AI with citizen science and participatory early warning systems. Future platforms will draw on user-generated data—from smartphone observations to wearable sensor alerts—to train and refine AI models. Local voices will shape environmental monitoring, with AI adapting to linguistic, cultural, and geographic contexts. In flood-prone countries like Nepal or the Philippines, AI systems are beginning to ingest multilingual crowd reports, SMS-based hazard updates, and drone footage to deliver timely, community-specific alerts (Google Crisis AI Report, 2024).

AI is also becoming deeply embedded in health surveillance systems linked to climate stress. With rising temperatures and disaster-related disruptions to public health infrastructure, AI models are increasingly used to predict disease outbreaks by analyzing climatic, sanitation, and health system data. Wearables that detect vital sign anomalies are linked to medical response dashboards, alerting first responders to heat stroke, respiratory crises, or clusters of waterborne illnesses. These health-climate AI linkages are already being piloted in cities like Houston, Mumbai, and Dakar (World Health AI & Climate Nexus Report, 2025).

In terms of physical response infrastructure, the coming years will see the rise of autonomous and AI-coordinated emergency fleets. These include firefighting drones, amphibious rescue robots, and medical delivery drones guided by swarm intelligence algorithms. AI-driven fleets self-organize in real-time, adjusting flight paths or deployment zones based on evolving threats, population movements, and terrain accessibility. This operational flexibility has been tested in California and British Columbia during wildfire responses in 2023–2024 (Natural Resources Canada, 2024).

Ethical governance and transparency will be a defining focus of AI deployment in emergency response. As AI systems make increasingly high-stakes decisions, calls for equitable and explainable AI are intensifying. Bias audits are being used to ensure that marginalized communities—including informal settlements, refugees, and Indigenous groups—are not deprioritized by algorithms. Regulatory frameworks are evolving to mandate publicly accessible AI decision criteria and equity-driven recalibration tools (Open Climate AI Ethics Consortium, 2025).

Globally, bioregional AI networks are emerging to manage transboundary climate risks. These networks, rather than being constrained by national borders, are designed to operate across ecosystems such as river basins, forest corridors, and mountain ranges. Data from upstream areas are being used to inform emergency preparedness in downstream areas. For example, glacial melt monitoring in the Himalayas is feeding predictive flood alerts in Bangladesh, while fire detection across the Amazon Basin supports coordinated action among Brazil, Peru, and Colombia (International Bioregional AI Partnership, 2025).

The financial and recovery sectors are also adopting AI tools. Insurance companies and governments are using AI to automate post-disaster damage verification and accelerate claims processing. AI algorithms trained on pre-disaster satellite images, facial recognition data, and geospatial damage assessments are being used to guide fairer, faster payouts. Some municipalities, including parts of Japan and Mexico, have begun using resilience scores generated by AI to prioritize rebuilding in safer zones (Munich Climate Risk Lab, 2024).

AI is also playing an essential role in the education and training of emergency responders. New simulation platforms using AI and augmented reality allow first responders to rehearse disaster scenarios with local variables, real-time feedback, and climate-specific threats. These immersive training environments are helping to shorten response times and improve coordination under pressure (IFRC Simulation & Tech Futures Report, 2025).

By 2035, it is projected that more than 120 countries will integrate AI into national disaster strategies, up from just over 45 in 2025 (UNDER, 2025).

The global market for AI disaster response tools is expected to grow from $5 billion in 2025 to over $30 billion by 2035 (Allied Market Research, 2024). Coverage of AI-powered early warning systems is forecast to expand from approximately 300 million people today to over 2 billion in the next decade. The time required for AI-assisted damage assessment will decrease from the current three to seven days to under 24 hours in most developed regions. Meanwhile, autonomous emergency fleets, currently in pilot stages, are expected to become standard practice in high-risk areas within the next ten years.

In conclusion, AI is not only revolutionizing emergency response but also reshaping humanity's approach to anticipating, managing, and recovering from climate-related disasters. The full potential of these technologies will be realized only if the principles of climate justice, equity, and collective resilience guide implementation.

D. A Case Study of Japan: A National Response

Japan has experienced many natural disasters. Japan's history is deeply intertwined with natural disasters due to its unique geological and climatic positioning. Located along the Pacific Ring of Fire, the country is highly seismically active, with thousands of earthquakes occurring each year. The earliest documented quake, the 416 CE Yamato earthquake, is recorded in ancient chronicles, and the 869 Jōgan tsunami devastated northeastern Japan with consequences eerily similar to the 2011 Great East Japan Earthquake. Volcanic eruptions also pose a persistent threat, with Mount Fuji and Mount Unzen among the most notable examples. Over centuries, these seismic events shaped Japan's architecture, spiritual practices, and communal response systems, embedding risk into the fabric of its society.

Typhoons have also played a dramatic role in Japanese history, often bringing catastrophic winds, floods, and landslides. The 1274 and 1281 typhoons, famously known as the "kamikaze" or "divine winds," repelled Mongol invasions and became symbols of national protection. In more recent history, Typhoon Isewan in 1959 left over 5,000 dead or missing, prompting sweeping changes to disaster management infrastructure.

Annual typhoons continue to pose a significant hazard, particularly in coastal and mountainous regions, and their intensity is increasing due to climate change, adding further urgency to Japan's adaptation strategies.

Japan's experience with disaster has led to the development of some of the most advanced emergency response systems in the world. Following the Kobe Earthquake in 1995, which resulted in over 6,000 deaths and exposed systemic failures in urban planning and government response, Japan revised its disaster laws and communication protocols. The 2011 Tōhoku earthquake, tsunami, and Fukushima nuclear disaster further emphasized the risks of cascading disasters and the need for resilient, decentralized infrastructure. Japan now invests heavily in public education, disaster drills, early-warning systems, and innovative urban planning, including natural disaster parks that double as community resilience hubs. These historical experiences continue to inform and inspire global models of preparedness and recovery.

- Japan faces over 1,500 earthquakes per year, and more than 70% of its land is disaster-prone.
- According to the Cabinet Office of Japan (2024), over 500 disaster prevention parks exist nationwide.

Japan's disaster parks are now studied and emulated worldwide, especially in:

- Chile (earthquake-prone urban areas)
- New Zealand (Wellington's tsunami evacuation green corridors)
- Indonesia (post-tsunami planning in Aceh)

A comparison between Japanese natural disaster parks and those in the US or Canada reveals significant differences in design philosophy, community integration, and urban planning goals. While Japan leads in integrating disaster preparedness directly into public space, the US and Canada tend to separate emergency planning from recreational or everyday landscapes. United States: Disjointed Emergency Infrastructure

In contrast, US cities typically separate parks from emergency infrastructure. Disaster preparedness is often the domain of agencies like FEMA, and urban parks are rarely designed with integrated emergency use in mind. Some cities have begun pilot projects:

- San Francisco's Resilient Corridors program explores how parks can serve as lifelines during earthquakes.
- Los Angeles has installed solar-powered charging benches and water stations in some parks but without full integration.
- New York City's flood resilience plans include stormwater-retaining green spaces, but these are climate mitigation, not disaster response parks.

The US approach is more agency-driven and infrastructure-heavy, often focused on post-disaster response rather than public-facing, integrated design.

Canada: Green Infrastructure for Climate Resilience
Canada incorporates green infrastructure and climate adaptation more proactively than the US but less so in terms of disaster-readiness embedded in parks. Examples include:

- Toronto's Corktown Common, which doubles as flood protection and public green space.
- Calgary's Bow River projects, integrating floodplain restoration into urban landscapes.
- However, dedicated evacuation or earthquake-resilient parks are still rare, especially outside seismically active zones.

Canada emphasizes climate resilience, biodiversity, and ecosystem restoration in public spaces rather than direct emergency preparedness.

Japanese disaster parks embody a holistic, culturally ingrained approach to integrating daily life with emergency preparedness. US and Canadian cities, while advancing in climate adaptation and green infrastructure, lag in creating multi-use disaster-ready public spaces. Adopting elements of Japan's approach could significantly improve community resilience in North America—especially as climate-driven disasters increase.

Japanese Natural Disaster Parks—also known as disaster prevention parks (防災公園, bōsai kōen)—are a distinctive part of Japan's approach to disaster resilience, risk communication, and community preparedness. These parks blend public green space with essential emergency infrastructure, education, and planning. Japan, being highly prone to earthquakes, tsunamis, typhoons, and volcanic eruptions, has pioneered this integrated model for urban resilience.

Disaster parks serve dual purposes:

1. Recreational Green Space during regular times.
2. Emergency Shelters and Operation Bases during disasters.

They are equipped with:

- Evacuation areas for thousands of people
- Helipads for rescue operations
- Water tanks, emergency toilets, and solar-powered lighting
- Emergency stockpiles (food, blankets, medical supplies)
- Underground cisterns for firefighting
- Education facilities and museums for disaster learning

Tokyo aims for a 30-minute walking distance to an emergency evacuation site for all residents.
Japanese Natural Disaster Parks are more than parks—they are living laboratories, shelters, education centers, and proof of how climate adaptation, urban design, and public education can be integrated. They embody Japan's commitment to living with nature rather than against it.

Climate Adaptation Guide
Here is a Community Adaptation Guide inspired by Japanese Natural Disaster Parks. This guide provides actionable steps for urban planners, educators, community leaders, and residents to implement disaster-resilient public spaces, drawing on Japan's model.

Community Adaptation Guide: Creating Disaster-Resilient Parks Inspired by Japan

1. Vision and Planning

To create a park that functions as both a green public space and a disaster-ready zone, communities must begin with a shared vision. This process starts by assessing local risks, including floods, fires, earthquakes, and extreme heat. Urban planners and residents should identify underused land or existing parks that could be redesigned for dual use. Planning should be participatory and inclusive, with workshops that invite input from residents, emergency managers, educators, and youth. Tools such as GIS mapping and local climate risk projections can inform these conversations. Ultimately, the goal is to co-create a plan that serves both recreation and emergency response.

2. Design Features for Resilience

The design of a disaster-resilient park requires infrastructure and landscaping that serve both everyday enjoyment and crises. Parks should include designated evacuation zones and open spaces large enough to gather people safely. Rainwater collection systems, solar panels with backup battery storage, and emergency supply pods should be integrated seamlessly into the landscape. Sanitation facilities such as dry or composting toilets must be made available, especially in regions where water access may be limited during an emergency. Educational and cultural elements are just as essential: parks can host signage about past disasters, include interactive simulations for children, and ensure all emergency instructions are accessible in multiple languages. Green infrastructure, such as native plantings, bioswales, and shaded areas utilizing tree canopies or solar panel structures, offers nature-based protection while enhancing daily use.

3. Governance and Participation

For a disaster park to remain effective, it must be supported by active governance and long-term community involvement. Local governments or neighborhood councils should establish a Community Resilience Committee that helps maintain and manage the park's emergency components. This group can coordinate with city agencies to ensure that emergency stockpiles are replenished and critical infrastructure, such as solar batteries or water tanks, remains functional. Schools, libraries, and cultural organizations should be enlisted to run education programs or host events. The governance model must be dynamic, allowing community members to take ownership of resilience efforts and ensuring continuity through regular training and leadership development.

4. Education and Outreach

Education is at the heart of the Japanese model, and communities looking to adapt this strategy should make public learning visible and engaging. Each year, resilience-themed events, such as fairs, drills, or open houses, should be held in the park. Students can visit for hands-on learning experiences, while youth groups can participate in simulations, scavenger hunts, or storytelling sessions about climate and disaster histories. Elders and long-term residents should be invited to share their experiences of past disasters and how they were overcome, offering a vital intergenerational link. Teaching kits can be created for local teachers to use in school, and multilingual guides should be provided to reach immigrant, refugee, and non-English-speaking populations.

5. Monitoring and Long-Term Impact

Over time, the success of a disaster-resilient park should be evaluated based on community participation, infrastructure maintenance, and improvements in emergency preparedness and response. Metrics may include the number of people attending drills, the accessibility of the park to vulnerable populations such as older people or individuals with disabilities, and how quickly the area can be activated during a disaster. Annual audits and feedback sessions—conducted through town hall meetings, mobile apps, or research partnerships with local universities—help improve performance and inform future investments. These parks must evolve in response to changing climate conditions and growing communities.
Case Example: Oakland, California

Inspired by Tokyo Rinkai Disaster Prevention Park, the city of Oakland transformed a former brownfield site into a solar-powered green space that now doubles as an evacuation and resilience center. It includes native, drought-tolerant landscaping, rainwater harvesting systems, an underground emergency supply bunker, multilingual evacuation signage, and a youth-run climate design lab.
Case Example: Toronto's Corktown Common

Built-in response to catastrophic flood risks, Corktown Common sits atop a 2.4-meter-high berm and can protect 500 acres of downtown Toronto. The park includes a wetland, recreational space, and backup water systems. It serves as an emergency staging ground while remaining a celebrated urban nature retreat.
Case Example: Houston's Buffalo Bayou Park

After devastating floods, Houston redesigned its park system to become part of the flood infrastructure. Buffalo Bayou Park includes trails, wetlands, bridges, and emergency access routes. It protects neighborhoods downstream and serves as a gathering space during normal conditions.
Case Example: Resilience Hubs in Baltimore and Miami

Baltimore's pilot resilience hub, located at the Madison Square Recreation Center, offers cooling, solar-powered charging, and preparedness training. Miami-Dade's network of resilience hubs includes parks that double as shelters, with an emphasis on social services and bilingual outreach.

This guide demonstrates how communities across North America can adapt Japanese disaster park models to their landscapes and risks. By linking environmental planning with education, emergency preparedness, and cultural inclusion, cities can turn parks into lifelines—before, during, and after disasters.

E. Conclusion

The role of emergency responders, the use of AI, the Japanese Case study and its application to the US, an advocacy brief, and a planner guide, all described above, will assist the climate adaptation generation in creating a viable blueprint for the future.

Advocacy Brief

Prepared by: [Your Name / Organization]
Date: June 2025
Executive Summary

The escalating impacts of climate change—extreme heat, intensified storms, wildfires, floods, and disease outbreaks—are overwhelming emergency response systems worldwide. Emergency responders are our first line of defense, yet they remain under-resourced, undertrained for climate-specific risks, and often excluded from long-term climate adaptation planning. This advocacy brief calls for urgent investment in climate-informed emergency response infrastructure, training, and equity-centered policy reform.

Problem Statement Climate change is not a future threat—it is a current emergency. In the past five years alone, the United States has experienced over $600 billion in climate-related disaster damages (NOAA, 2025), while globally, extreme weather has displaced more than 40 million people annually (IDMC, 2024). Yet, emergency response systems remain reactive, fragmented, and vulnerable to being overwhelmed by the scale, frequency, and complexity of climate-related disasters.

Key Challenges

1. Outdated Response Models: Most systems are designed for short-term disasters rather than long-duration, compound climate crises (e.g., wildfires followed by flooding or disease outbreaks).
2. Lack of Climate Training: Emergency responders often lack standardized, climate-specific education, particularly in interpreting predictive models and assessing ecological risks.

3. *Underinvestment in Rural and Marginalized Areas:* Disadvantaged communities face longer emergency response times, fewer resources, and lower survival rates.
4. *Mental Health and Workforce Burnout:* First responders are experiencing increasing rates of PTSD and burnout due to repeated climate trauma without adequate psychological support systems.
5. *Technological Disparities:* AI, drones, and data tools for disaster management are not equitably distributed across regions or agencies.

Policy Recommendations

1. Federal and State Climate Resilience Funding
- *Establish a Climate Emergency Response Innovation Fund to equip local agencies with resilient infrastructure, mobile technologies, and clean backup power systems.*
- *Prioritize funding for rural, tribal, and underserved communities.*

2. Mandatory Climate Adaptation Training
- *Integrate climate risk modules into FEMA, paramedic, fire, and EMS training curricula.*
- *Provide certification programs in climate-informed emergency response for local, state, and private responders.*

3. Public-Responder Partnerships
- *Launch Community Resilience Hubs, where responders conduct outreach, drills, and education on heat waves, floods, wildfire evacuation, and chronic climate risks.*

4. AI and Data-Driven Risk Tools
- *Support open-source development and deployment of AI early warning systems, heat vulnerability maps, and dynamic evacuation platforms.*
- *Promote public access to real-time disaster forecasts, especially in Indigenous, immigrant, and low-income neighborhoods.*

5. Mental Health Services for Responders
- *Embed climate grief and trauma support into responder wellness programs.*
- *Fund on-scene mental health teams during major climate events to assist both the public and responders.*

Urgency and Opportunity

We face a narrow window to transform emergency response systems from reactive to adaptive. Climate-related disasters are accelerating, yet responders remain our most trusted and locally embedded line of defense. With federal coordination, state innovation, and local empowerment, we can develop a modern emergency response infrastructure that not only saves lives but also fosters long-term community resilience.

Call to Action

We urge legislators, municipal leaders, emergency response agencies, and philanthropic partners to:

- *Co-sponsor and fund the Emergency Responders Climate Readiness Act (ERCRA).*
- *Integrate emergency services into national and regional Climate Adaptation Plans.*
- *Mandate inclusion of responder voices in climate policy development.*

For every dollar invested in resilient disaster response, $4–$8 is saved in avoided damage and recovery costs (World Bank, 2024). Climate adaptation must begin with those who respond first—and stay last.

3Prepared by:
[Your Name / Organization]
Contact: [Email / Phone]
LinkedIn: [Insert]
Website: [Insert]

Planning Toolkit

Title: Climate-Resilient Emergency Response Planning Toolkit

Audience: Municipal planners, resilience officers, emergency managers, land-use agencies

Date: June 2025

Prepared by: [Your Name / Organization]

Scope: Urban, rural, and bioregional planning contexts

Toolkit Objectives

1. *Integrate emergency response systems into climate adaptation planning.*
2. *Provide data-driven, equitable, and actionable strategies.*
3. *Enhance coordination between planners, first responders, public health, and infrastructure sectors.*
4. *Align planning with the IPCC AR6, FEMA BRIC, and national and local climate resilience frameworks.*

Section 1: Climate Risk + Emergency Response Assessment Tools

A. Multi-Hazard Climate Mapping

- *Use downscaled regional climate projections to map:*

- *Wildfire corridors*
- *Sea-level rise/floodplains*
- *Extreme heat and heat island zones*
- *Earthquake fault lines*
- *Air pollution and disease vector hotspots*

- *AI platforms: Google FloodHub, IBM Geospatial AI*

B. Emergency Infrastructure Vulnerability Audit

Conduct audits of:

- *Fire stations, hospitals, dispatch centers*
- *Cooling/warming centers*
- *Evacuation routes and emergency signage*
- *Back-up power, water, and communication systems*

Checklist Includes:

☐ *Located outside high-risk zones*
☐ *Passive design for heat/flood events*
☐ *Solar microgrid capacity*
☐ *ADA compliance and language accessibility*
☐ *Accessible to frontline and high-vulnerability populations*

Section 2: Land Use & Zoning Adaptation

A. Code and Ordinance Updates

- *Update zoning to prohibit high-density development in flood/fire-prone areas.*
- *Require emergency response access roads in all new subdivisions.*
- *Mandate Emergency Climate Impact Statements (ECIS) for significant developments.*

B. Green Infrastructure for Resilience

- *Include fire breaks, wetlands, bioswales, and stormwater parks.*
- *Dual-use design: parks as flood retention + evacuation assembly zones.*

Section 3: Equitable Community Preparedness Design

A. Resilience Hubs & Neighborhood Response Zones

- *Identify high-risk zones by combining:*
  - *CalEnviroScreen, CDC SVI, tribal consultation*
  - *Public health and social vulnerability data*
- *Designate trusted community locations for response coordination (e.g., schools, mosques, libraries).*
- *Include multilingual signage, food storage, water purification systems, and mental health first aid kits.*

B. Mobility and Evacuation Planning

- *Create shaded, ADA-accessible evacuation routes.*
- *Utilize predictive AI to identify bottlenecks and gaps in first-mile access.*
- *Partner with transit agencies for paratransit surge capacity.*

Section 4: Capacity Building and Workforce Resilience

A. Interdisciplinary Training Modules

- *For Planners: Climate risk assessment, trauma-informed design, AI tools, equity integration.*
- *For Responders: Climate Literacy, Ecosystem-Informed Evacuation, and Indigenous Consultation Protocols.*

Training Platforms:

- *FEMA Emergency Management Institute*
- *National League of Cities Climate Resilience Courses*
- *AI4Climate Workshop Series (2024–2025)*

B. Climate Response Drills and Exercises

- *Conduct annual compound climate crisis simulations (e.g., wildfire + blackout + landslide).*
- *Include schools, hospitals, utilities, and social service partners.*
- *Debrief and revise municipal adaptation plans following the drill.*

Next Steps

- *Customize this toolkit to your region's climate risk profile and planning authority.*
- *Host a Climate-Ready Emergency Summit with local stakeholders and responders.*
- *Report progress annually and publish post-event adaptation lessons learned.*

Contact for Support and Technical Assistance:

[Name / Agency]
[Email] | [Phone] | [Website]

Resources

1. Allied Market Research. (2024). *AI for disaster response: Market forecast 2025–2035*. https://www.alliedmarketresearch.com
2. Cruz, M. J., & Kumar, S. (2024). Artificial intelligence in climate adaptation and disaster resilience. *Climate Intelligence Review*, 11(2), 45–61.
3. FEMA AI Futures Lab. (2025). *Integrating machine learning in federal disaster planning*. US Department of Homeland Security.
4. Google Crisis AI Report. (2024). *Flood Forecasting and AI-Powered Alerts: Lessons from South Asia*. Google Research.
5. IFRC Simulation & Tech Futures Report. (2025). *Next-gen emergency responder training with AI and VR*. International Federation of Red Cross and Red Crescent Societies.
6. International Bioregional AI Partnership. (2025). *Cross-border AI solutions for climate emergencies in shared ecosystems*. Global Climate Response Network.
7. Munich Climate Risk Lab. (2024). *AI and insurance: Redesigning risk in the era of climate volatility*. Munich Re.
8. Natural Resources Canada. (2024). *AI-guided wildfire management in Canada: 2023 pilot summary*. Government of Canada.
9. Open Climate AI Ethics Consortium. (2025). *Equity and Transparency in AI Disaster Systems*. https://www.climateaiconsortium.org
10. United Nations Office for Disaster Risk Reduction (UNDRR). (2025). *Harnessing AI for disaster risk reduction and resilience*. https://www.undrr.org
11. World Bank. (2024). *AI and Edge Computing for Inclusive Climate Risk Management in the Global South*. https://www.worldbank.org
12. World Health AI & Climate Nexus Report. (2025). *Artificial intelligence and global health resilience to climate shocks*. WHO & AI for Health Alliance.
13. World Bank. (2024). *AI for disaster risk management: Case studies from Asia and Africa*. https://www.worldbank.org
14. Tokyo Metropolitan Government. (2024). *Disaster Prevention Park Guidelines*.
15. United Nations Office for Disaster Risk Reduction (UNDRR). (2025). *Nature-Based Solutions for Urban Risk Reduction*.
16. Ministry of Land, Infrastructure, Transport and Tourism Japan. (2024). *Integrated Green Infrastructure Strategies for Disaster Resilience*.

17. ICLEI Resilient Cities Network. (2025). *Community-Led Adaptation Tools and Resources.*
18. The city of Houston. (2025). *Buffalo Bayou Park: Green Infrastructure in a Changing Climate.*
19. Resilient Cities Network. (2024). *North American Resilience Hubs Toolkit.*
20. City of Toronto. (2025). *Corktown Common Flood Protection and Urban Design Report.*
21. US Climate Resilience Toolkit (NOAA). (2024). *Community Adaptation Planning Case Studies.*
22. ICLEI USA. (2024). *Designing for Resilience in Public Space: A Municipal Guide.*
23. National League of Cities. (2025). *Equity and Emergency Management: Parks as Resilience Infrastructure.*

Chapter Three
Floods

Ban Ki-moon, former UN Secretary-General:

"We are the first generation that can end poverty, and the last that can take steps to avoid the worst impacts of climate change. Future generations will judge us harshly if we fail to uphold our moral and historical responsibilities."

Flood-related disasters have escalated dramatically over the past two decades. According to the World Meteorological Organization (2024), such events increased globally by 134% between 2000 and 2023. The United Nations Office for Disaster Risk Reduction (UNDRR) reports that over 1.5 billion people now live in flood-prone regions, and that flooding accounts for 90% of all climate-related disasters worldwide. Financial costs are also soaring; the Swiss Re Institute (2024) recorded global insured losses from flooding at $60 billion USD in 2023 alone, reflecting both the growing frequency and severity of these events. These figures underscore the magnitude of the challenge and the importance of investing in forward-looking adaptation measures.

Adaptation to flooding involves proactive and reactive measures that reduce the vulnerability of communities, ecosystems, and infrastructure to flood-related impacts, especially as climate change increases the frequency and severity of such events.

Flooding takes several distinct forms, each shaped by environmental, climatic, and urban conditions. Riverine flooding occurs when rivers overflow their banks, often due to prolonged rainfall or snowmelt that exceeds the capacity of the river channel. Urban flooding, by contrast, is a direct consequence of impervious surfaces in city landscapes—such as concrete and asphalt—which prevent water absorption and overwhelm aging or insufficient drainage systems. Coastal flooding is driven by sea-level rise, storm surges, and tropical cyclones, with low-lying coastal zones and island nations particularly at risk.

Flash flooding is the most sudden and dangerous type, resulting from intense rainfall, dam breaches, or rapid snowmelt, often with little warning time for evacuation.

A. Adaptation Strategies: Categories

Adaptation strategies to manage and reduce flood risk span five major categories. The first involves structural measures that engineer physical barriers and storage systems to control floodwaters. Examples include levees, dikes, and floodwalls, such as the sophisticated Delta Works in the Netherlands that protect urban and agricultural areas from rising waters. Detention basins and reservoirs are also vital in temporarily holding excess water, thereby mitigating downstream inundation. In urban areas, green infrastructure, including permeable pavements, rain gardens, bioswales, and green roofs—plays an increasingly important role in reducing surface runoff and enhancing water absorption.

A second category includes nature-based solutions that restore or leverage ecological systems to absorb or deflect floodwaters. Wetland restoration, for example, acts as a sponge for excess water while also enhancing biodiversity. Coastal mangrove forests serve as natural buffers against storm surges and erosion; a strategy effectively deployed in nations like the Philippines and Bangladesh. Reforestation in upland regions improves watershed stability and reduces the intensity of flash flooding through enhanced soil retention and infiltration.

The third adaptation area centers on land use planning, which seeks to reduce exposure by regulating how and where development occurs. Zoning laws that prohibit construction in flood-prone zones are foundational to long-term risk reduction. In some high-risk areas, managed retreat—relocating people and infrastructure away from vulnerable locations—has been implemented, as seen in New York's Staten Island buyout program. Accurate floodplain mapping and risk zoning are essential tools in guiding both building codes and flood insurance frameworks.

A fourth line of defense involves early warning systems and emergency response. Technological platforms such as the European Flood Awareness System provide real-time forecasting using advanced modeling and meteorological data. Locally, communities benefit from alert mechanisms like SMS warnings, sirens, and mobile applications that disseminate urgent flood information. Pre-planned evacuation routes and regular disaster drills ensure that people know how to respond quickly, reducing casualties and chaos during flood events.

Finally, socioeconomic and institutional measures support broader resilience goals. Flood insurance and risk-transfer mechanisms—through public programs or private markets—help communities recover financially after flood events. Equally critical is public education: raising awareness about flood risks and encouraging households to prepare through actions like securing valuables, reinforcing structures, and maintaining emergency kits. At the systemic level, Integrated Water Resource Management (IWRM) promotes coordinated planning for water, land, and environmental resources to ensure long-term sustainability and reduce flood vulnerability across sectors.

B. Examples from Around the World

Region:	Example Adaptation Project
Netherlands:	Delta Program: combines sea defenses, spatial planning, and water storage.
Bangladesh:	Raised homesteads, floating agriculture, and community warning systems.
U.S. Gulf Coast:	Wetland restoration and home elevation programs in Louisiana.
Jakarta, Indonesia:	Giant seawall project, plus planned city relocation to Borneo.
Canada:	High River, Alberta: post-2013 flood redesign of neighborhoods and infrastructure.

C. Challenges to Adaptation

Adapting to climate-induced flooding faces several significant hurdles that stem from structural inequities, aging infrastructure, and uncertainty in climate modeling. One of the most pressing challenges is equity and displacement. Low-income and marginalized communities are often located in high-risk flood zones and lack the financial means to relocate, retrofit homes, or access insurance. These groups may face forced displacement without adequate compensation or relocation support, deepening cycles of poverty and vulnerability.

Another challenge is infrastructure limitations. Many cities and towns—especially in the Global North—rely on outdated or undersized drainage and sewer systems built for past climate conditions. As rainfall intensity and frequency increases, these systems are quickly overwhelmed, leading to frequent flooding and associated public health risks.

Policy gaps further hinder adaptation. A lack of coordination across local, regional, and national levels often results in fragmented or inconsistent planning, making it difficult to implement effective flood risk reduction strategies. This is compounded by climate uncertainty, particularly around localized precipitation patterns, which makes it difficult for planners and engineers to accurately design for future conditions.

D. Future Directions

To address these multifaceted challenges, several forward-looking strategies are emerging. Smart flood management systems are being deployed in cities like Singapore, where artificial intelligence (AI) and the Internet of Things (IoT) enable real-time flood forecasting, dynamic control of drainage infrastructure, and targeted emergency response.

Investments in climate-resilient infrastructure are becoming more critical. This involves designing bridges, roadways, public buildings, and transit systems with flood-resistant materials, elevated foundations, and passive drainage systems that can endure and recover from severe inundation.

Adaptive governance offers a promising model for the future. Instead of static, one-time plans, governance structures must remain flexible and iterative, allowing cities to recalibrate strategies as flood risks evolve. This includes revisiting zoning laws, updating building codes, and adjusting insurance frameworks based on real-time climate data.

Equally important is community-led adaptation, which places decision-making power in the hands of those most affected. Providing local communities with the funding, training, and institutional support to initiate their own flood mitigation strategies—such as neighborhood rain gardens, localized warning systems, or participatory mapping—builds long-term resilience and fosters environmental justice.

E. Adaptation to Flooding: Specific Strategies, Examples, and Challenges

Adaptation to flooding involves a range of strategies designed to reduce the vulnerability of human and ecological systems to water-related disasters. As climate change intensifies precipitation extremes, sea level rise, and the frequency of catastrophic storms, the urgency to adapt to flooding grows across every continent. Flooding can take multiple forms, including riverine overflow due to heavy rainfall or snowmelt, urban flooding caused by inadequate drainage systems, coastal flooding resulting from storm surges and sea level rise, and flash flooding from sudden intense downpours or dam failures.

One major adaptation category involves structural interventions. These include the construction of levees, dikes, and floodwalls to protect urban centers and agricultural land from encroaching water. Detention basins and reservoirs are designed to temporarily store excess water to prevent downstream overflow. Increasingly, cities are turning to green infrastructure such as permeable pavement, bioswales, rain gardens, and vegetated rooftops to absorb runoff and slow stormwater. These nature-compatible approaches are especially useful in dense urban environments.

Nature-based solutions are also proving vital in reducing flood risk while offering co-benefits to ecosystems. Restoring wetlands provides space for water to spread during flood events while enhancing biodiversity. In coastal zones, mangrove forests serve as living barriers that absorb wave energy and reduce storm surge impacts. Reforestation in upland areas helps improve watershed management and reduces the risk of landslides and flash floods by increasing soil stability and water absorption.

Land use planning remains a critical tool in long-term flood adaptation. Some governments restrict new construction in flood-prone zones or elevate building codes to require higher floor elevations. Managed retreat is another approach, involving the voluntary relocation of homes and infrastructure away from high-risk areas. One example is New York City's Staten Island retreat program following Hurricane Sandy. Floodplain mapping and zoning laws help guide these decisions, and they are often supported by updated climate and hydrological data.

Early warning systems and emergency response are essential elements of adaptive flood management. Forecasting systems now use real-time satellite and sensor data to predict when and where floods will occur. Community-based alert systems using SMS, sirens, and mobile apps help inform populations quickly. Evacuation drills and local response planning are integrated into school curricula and community risk reduction programs, especially in regions with frequent flooding.

Socioeconomic and institutional approaches also play a key role. Flood insurance, both public and private, can help individuals recover from disaster, although access is often unequal. Community education campaigns aim to increase awareness of local flood risks and teach household preparedness, including how to secure valuables, create emergency kits, and know safe evacuation routes.

Integrated Water Resource Management (IWRM) frameworks support coordinated planning between water, land, and disaster management agencies, ensuring that flood policies align with climate adaptation goals.

Numerous countries are pioneering innovative flood adaptation programs. The Netherlands' Delta Program is one of the world's most comprehensive, combining sea defenses, spatial planning, and long-term climate resilience. In Bangladesh, floating agriculture, elevated homes, and cyclone shelters reduce the risks posed by seasonal floods. In the United States, Louisiana has elevated entire neighborhoods, and in Canada, the town of High River, Alberta, redesigned its infrastructure after devastating floods in 2013. Jakarta, Indonesia, is planning a massive seawall and considering relocating its capital due to chronic urban flooding. These examples demonstrate the range of tools being applied across different geographies and income levels.

Despite advances, adaptation to flooding faces serious challenges. Social inequities are often deepened, as low-income and marginalized communities typically have fewer resources to retrofit homes or relocate. Aging infrastructure and underinvestment leave many cities vulnerable to even moderate flooding. Policies can be fragmented across local, regional, and national levels, hampering coordinated efforts. Moreover, future rainfall and storm intensity remain difficult to predict at fine scales, making it hard to plan infrastructure with sufficient resilience.

Looking forward, flood management is evolving through technological innovation and participatory governance. Smart flood systems use artificial intelligence, sensors, and real-time modeling to optimize drainage and emergency response, as seen in Singapore's Smart Drainage System. New standards are emerging for climate-resilient infrastructure, requiring roads, bridges, and public buildings to withstand future flood events. Adaptive governance frameworks aim to make policies flexible and responsive to evolving risks, while community-led adaptation efforts place decision-making power in the hands of those most directly affected.

Recent research and global reports provide key insights into effective flood adaptation. The UN Office for Disaster Risk Reduction's 2024 Global Assessment Report outlines best practices in community resilience. The World Meteorological Organization's 2024 State of Global Climate Report tracks global precipitation and flood trends. Swiss Re's 2024 Global Catastrophe Losses Report provides economic data on flood-related damages.

Peer-reviewed work such as Aerts and Botzen's 2025 article in Climate Policy Journal analyzes evolving urban flood risk strategies. The IPCC's Sixth Assessment Report synthesizes the latest climate science related to hydrological extremes and flood forecasting.

F. Global Focus: Flooding in the IPCC Sixth Assessment Report (AR6)

The Intergovernmental Panel on Climate Change (IPCC) in its Sixth Assessment Report (AR6) presents conclusive and urgent evidence that flood risks are rising globally due to human-induced climate change. Flooding is identified as one of the most severe and widespread climate-related hazards, impacting billions of people, especially in densely populated and low-lying areas. The report underscores that without aggressive adaptation and mitigation strategies, the magnitude, frequency, and economic cost of flooding will escalate dramatically in the coming decades.

In the Physical Science Basis report (Working Group I), the IPCC confirms with high confidence that increased atmospheric moisture caused by global warming is directly intensifying heavy precipitation events. For every 1°C of global temperature rise, the atmosphere holds approximately 7% more water vapor, fueling more intense rainfall.

The IPCC attributes the growing intensity and frequency of extreme rainfall events, and the increased occurrence of compound flooding—such as the simultaneous overflow of rivers and storm surges in coastal areas—to anthropogenic climate change. Regions such as Central and Western Europe, South Asia, East Asia, and coastal areas of North America have experienced a marked increase in such events. Glacier-fed rivers in the Himalayas and Andes are increasingly vulnerable to flooding due to accelerated glacial melt and the threat of glacial lake outburst floods. Rapid urbanization, especially in megacities, has exacerbated pluvial (surface water) flooding due to paved surfaces and inadequate drainage systems.

The report on Impacts, Adaptation, and Vulnerability (Working Group II) highlights the social, ecological, and economic consequences of flooding. According to AR6, over 1.8 billion people, nearly one-quarter of the global population, are already exposed to significant flood risk. Economic losses from flood events have more than doubled since the 1990s, and without urgent intervention, they are projected to increase by 50 to 150 percent by 2050 under moderate emissions scenarios. In Asia and Africa, where urban growth often outpaces infrastructure, adaptation gaps are particularly wide.

Informal settlements in river basins and along coasts face disproportionate risks. The report also stresses the intensifying threat of compound events, such as flooding that occurs during pandemics or is followed by heatwaves, which places additional strain on health systems and emergency services. The concept of "limits to adaptation" is introduced, warning that some areas—such as parts of the Mekong Delta, low-lying Pacific Island nations, and river deltas in Bangladesh—may become uninhabitable if global warming exceeds 2°C, regardless of current adaptation technologies. The IPCC affirms that Indigenous knowledge and local adaptation strategies are essential to coping with flooding, particularly in regions where formal flood defenses are limited or absent.

In its Mitigation of Climate Change report (Working Group III), the IPCC links land use, urban development, and flood vulnerability. It emphasizes that poorly planned urban expansion and infrastructure decisions can lock communities into decades of flood risk. The report recommends integrating flood mitigation into climate-resilient urban design through strategies such as decentralized water storage, nature-based solutions, and stormwater management.

Urban planning that incorporates wetlands, parks, green roofs, and permeable surfaces can mitigate flooding while providing co-benefits for air quality and public health. The report stresses the need for policies that address both emissions reduction and disaster resilience, making a strong case for integrative development frameworks.

The AR6 synthesis report delivers several urgent messages. It states unequivocally that flood risks are growing because of climate change. Without rapid and equitable adaptation, human suffering and economic damage from flooding will continue to rise. Coastal megacities such as Jakarta, Dhaka, Lagos, and New York face mounting threats from multiple types of flooding, including sea level rise, storm surge, and intense rainfall. The report warns that the window for effective adaptation is narrowing and that failure to act now will result in escalating losses, irreversible damage to ecosystems, and widening inequality.

The 2023 synthesis update reinforces these findings with recent data. Sea levels have risen approximately 20 centimeters since 1900 and are accelerating. Under high-emission scenarios, global sea levels could rise by as much as 1.1 meters by the end of the century, amplifying coastal flood risks and inundating critical infrastructure. Projections suggest that by 2050, over 570 coastal cities will suffer annual flood-related damages exceeding one trillion U.S. dollars unless large-scale adaptation investments are made.

In a world with 2°C of warming, events that are now considered once-in-a-century floods could occur every 5 to 10 years in many regions, fundamentally altering how societies must prepare for and manage risk.

The IPCC AR6 stresses that adaptation to flooding must be immediate, coordinated, and transformational. This includes the rapid deployment of structural and non-structural adaptation strategies, the integration of local and Indigenous knowledge, and long-term planning that incorporates flood risk into economic, social, and spatial policies. The report calls for governments, businesses, and communities to act with urgency, recognizing that the costs of inaction far outweigh the investments required to build flood-resilient societies. Without sustained global commitment, flood risks will continue to endanger the health, livelihoods, and stability of billions of people.

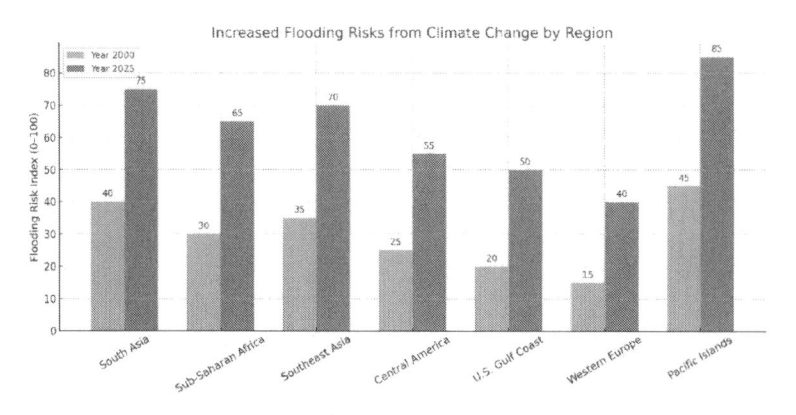

Increased Flooding Risks from Climate Change by Region

Region	Year 2000	Year 2025
South Asia	40	75
Sub-Saharan Africa	30	65
Southeast Asia	35	70
Central America	25	55
U.S. Gulf Coast	20	50
Western Europe	15	40
Pacific Islands	45	85

G. The Role of Artificial Intelligence

Artificial intelligence is rapidly reshaping how we manage climate-induced flooding, bringing new levels of precision, speed, and scale to one of the planet's most devastating hazards. A growing number of AI-powered platforms are now in use—from early warning systems in Asia to real-time crisis mapping in North America, each designed to anticipate, mitigate, and respond to flooding events driven by climate change.

One of the most globally recognized initiatives is Google's FloodHub, an early AI-driven warning system that uses hydrological models, satellite data, and machine learning to predict riverine floods in high-risk regions. Launched initially in India and Bangladesh, where millions live along floodplains, the system now covers over 80 countries as of 2025. Google's platform delivers hyperlocal alerts days in advance via mobile phones and community partners, proving vital for disaster preparedness in low-income regions.

In Southeast Asia, Singapore's Smart Drainage System integrates artificial intelligence with IoT sensors to manage stormwater dynamically. Using real-time rainfall and flow data, the system redirects water through canals and retention basins to prevent flash floods in urban neighborhoods. AI algorithms predict peak flow locations and adjust the system automatically, optimizing public safety and reducing infrastructure strain.

The IBM Green Horizon Project has developed AI models for flood risk analysis in Chinese megacities. By combining historical flood records, climate projections, and urban growth models, the system forecasts water surge risks under various scenarios. This helps planners redesign city layouts and infrastructure placement to avoid future disaster zones.

In Canada, Natural Resources Canada's AI-guided Flood Mapping Program uses convolutional neural networks to interpret satellite imagery and detect flood extents across large landscapes. These AI-generated maps are significantly faster and more accurate than traditional models, making them invaluable for emergency responders during seasonal floods, especially in Indigenous and remote communities.

The European Space Agency's AI4EO Flood Project combines Earth observation data with AI to enhance continental flood monitoring. This system is part of the Copernicus Emergency Management Service and supports EU countries with near-real-time flood extent data, economic damage forecasting, and emergency logistics planning. Its open-source architecture allows academic and municipal partners to co-develop applications.

For community-level engagement, Flood AI by Cloud to Street uses satellite data and AI to offer localized flood risk assessments for governments and insurers. By providing real-time dashboards, risk modeling, and population exposure tools, it helps decision-makers prioritize infrastructure investment and humanitarian aid. The platform is particularly focused on Global South cities where flood data is scarce.

In the United States, FEMA's AI Futures Lab is piloting integration of machine learning into disaster planning models. These tools assist in simulating compound events, such as hurricane-induced flooding followed by storm surge, and prioritizing resource distribution across counties based on vulnerability indexes.

Lastly, DHI MIKE AI Flood Forecasting Suite, used in over 30 countries, merges AI with hydrodynamic modeling to improve predictive flood forecasting in rivers and urban drainage systems. The system learns over time from new hydrological events, improving both short- and long-term forecasts.

Together, these AI platforms are not just tools, they are critical infrastructures for climate adaptation. By combining predictive analytics, community-based alerting, and planning support, they are helping vulnerable populations and urban systems prepare for increasingly severe and frequent flooding.

Advocacy Brief

Executive Summary

Flooding is the most common and costly climate-related disaster worldwide, affecting over 1.5 billion people and accounting for 90% of all climate disasters (UNDRR, 2024). As extreme weather intensifies due to climate change, flood-related losses are escalating—reaching $60 billion USD in insured damages in 2023 alone (Swiss Re, 2024). This advocacy brief outlines urgent, evidence-based strategies to improve flood resilience through equitable, adaptive, and community-centered planning.

Problem Statement

Climate-induced flooding is increasing in frequency, severity, and unpredictability. Vulnerable populations—particularly low-income, Indigenous, and historically marginalized communities—are disproportionately impacted due to systemic inequities, outdated infrastructure, and policy fragmentation. Without immediate and sustained adaptation, the social, economic, and ecological toll will deepen.

Policy and Planning Challenges

- *Equity Gaps: Displacement and loss disproportionately affect communities with the fewest resources to prepare or recover.*
- *Aging Infrastructure: Drainage systems, levees, and transportation networks in many cities were not designed for current or future rainfall patterns.*
- *Lack of Coordinated Governance: Siloed policies between municipal, regional, and national agencies undermine integrated flood risk management.*
- *Climate Uncertainty: Downscaled precipitation models are limited, making precise local planning difficult.*

Recommended Actions

1. Invest in Climate-Resilient Infrastructure

Retrofit and redesign bridges, roads, and stormwater systems to withstand flooding. Prioritize green infrastructure such as bioswales, wetlands, and permeable surfaces that mitigate runoff while enhancing biodiversity.

2. Deploy Smart Flood Management Systems

Integrate artificial intelligence (AI), satellite monitoring, and sensor-based Internet of Things (IoT) systems to enable real-time forecasting, early warning, and dynamic response (e.g., Singapore's Smart Drainage System).

3. Adopt Adaptive Governance Models

Mandate flexible planning approaches that incorporate climate uncertainty, iterative assessments, and multi-jurisdictional collaboration. Update zoning codes and insurance frameworks to reflect future flood risks.

4. Ensure Equitable and Community-Led Adaptation

Support participatory planning that empowers local communities, particularly those at highest risk, with funding, technical tools, and policy decision-making authority. Invest in relocation assistance and retrofitting for at-risk housing.

5. Enforce Integrated Land Use Planning

Restrict development in floodplains, restore wetlands, and require Personal Environmental Impact Statements (PEIS) in flood-prone real estate projects. Implement cumulative impact assessments that consider public health, ecosystem loss, and displacement risks.

Key Statistics

- *Flood disasters increased 134% globally between 2000 and 2023 (WMO, 2024).*
- *1.5 billion people live in flood-prone areas, most of whom reside in the Global South (UNDRR, 2024).*
- *2023 saw $60 billion in insured flood losses globally (Swiss Re, 2024).*

Call to Action

We call on governments, regional planners, and international development agencies to:

- *Allocate climate adaptation funding to high-risk, under-resourced communities.*
- *Mandate integrated flood risk management in urban planning legislation.*
- *Institutionalize community-led adaptation practices in federal and municipal frameworks.*
- *Scale open-source, AI-based flood prediction tools to the Global South.*

Delaying adaptation will only raise the cost in human lives, public health crises, economic loss, and social fragmentation. The time to act is now.

Planner's Toolkit

Prepared for: Municipal Planning Departments, Regional Authorities, and Climate Resilience Practitioners
Date: June 2025

1. Risk Assessment and Mapping Tools

- *Flood Vulnerability Maps: Use high-resolution GIS, LiDAR, and satellite imagery to identify flood zones, infrastructure at risk, and vulnerable populations.*
  Example tools: FEMA Flood Map Service Center, Copernicus Emergency Management Service, Canada's Flood Mapping Framework.
- *Cumulative Impact Assessment: Evaluate overlapping stressors—climate, land use, socioeconomics, public health—to inform site-specific decisions.*
- *Social Vulnerability Index (SVI): Incorporate demographic data (age, income, language, mobility) to prioritize equity in adaptation planning.*

2. Climate-Proof Infrastructure Design

Green Infrastructure:

- *Permeable Pavements, bioswales, green roofs, and urban wetlands absorb runoff and reduce flood peaks.*
- *Mandate green infrastructure in building codes and development permits.*

Elevated and Floodable Design:

- *Elevate homes, utilities, and critical assets in flood-prone areas.*
- *Design parks, parking lots, and open spaces to absorb and redirect floodwaters safely.*

Resilient Retrofitting:

- *Strengthening existing stormwater systems with backflow preventers, overflow reservoirs, and smart drainage controls.*
- *Retrofit public buildings with water-resistant materials and raised foundations.*

3. Smart Monitoring and Early Warning Systems

AI-Based Forecasting Platforms: Deploy machine learning to integrate rainfall, tide, and river data for hyperlocal flood alerts.

- *Example: Singapore's Smart Drainage System; Google FloodHub.*

Sensor Networks and IoT Devices:

- *Install sensors in rivers, drains, and urban hotspots to detect water levels and trigger automated alerts.*

Mobile Alerts and Citizen Engagement:

- *Use multilingual, geotargeted alerts and apps for evacuations and preparedness tips.*

4. Policy and Regulatory Instruments

Updated Zoning and Land Use Regulations:

- *Ban or limit development in high-risk floodplains.*
- *Enforce setback zones, elevation requirements, and open-space preservation buffers.*

Personal Environmental Impact Statements (PEIS):

- *Require all real estate and infrastructure developments to assess climate, health, and cumulative impacts.*

Flood Insurance and Financial Incentives:

- *Promote community-wide coverage through the National Flood Insurance Program (NFIP) or equivalents.*
- *Provide grants or tax incentives for flood-proofing and elevation.*

5. Community-Led and Equitable Adaptation

Participatory Visioning and Planning:

- *Co-create solutions with at-risk residents through workshops, design charrettes, and local advisory councils.*

Relocation Support and Housing Justice:

- *Offer equitable buyout programs, rent subsidies, and relocation grants for displaced families.*

Youth and School-Based Engagement:

- *Integrate flood literacy and adaptation into K–12 education.*
- *Support school-based green infrastructure and student-led projects.*

6. Emergency Response and Recovery Planning

Integrated Flood Response Plans:

- *Align local emergency protocols with climate risk projections.*
- *Coordinate with fire, police, EMS, and utility operators in real-time.*

Flood-Resilient Shelters and Transport:

- *Pre-designate and retrofit shelters with backup power, elevated access, and sanitary facilities.*
- *Map emergency routes and install flood-resistant public transit systems.*

7. Funding and Implementation Resources

Federal and International Grants:

- *Tap into funding from FEMA BRIC, HUD CDBG-DR, World Bank Climate Adaptation Fund, and UNEP's Adaptation Gap Program.*

Public–Private Partnerships (PPPs):

- *Incentivize developers, insurers, and engineering firms to co-invest in climate-resilient infrastructure.*

Climate Bonds and Resilience Financing:

- *Issue municipal green bonds specifically for flood mitigation projects.*

8. Evaluation and Feedback Mechanisms

Resilience Metrics and Indicators:

- *Track flood frequency, economic losses, social dislocation, and infrastructure failure rates post-adaptation.*

Community Report Cards:

- *Use participatory tools for residents to assess preparedness, satisfaction, and equity in flood responses.*

Adaptive Policy Loops:

- *Schedule reviews every 3–5 years to revise plans based on new climate data and community input.*

Featured Case Studies

- *New York City, USA: Green Infrastructure Plan reduces runoff by 1.5 billion gallons annually.*
- *Rotterdam, Netherlands: "Water Squares" double as public parks and floodwater retention basins.*
- *Bangladesh: Community-based early warning systems and floating schools in flood-prone deltas.*

Resources

1. Aerts, J. C. J. H., & Botzen, W. (2025). Adaptation to flood risk in urban areas: New directions in policy and practice. *Climate Policy Journal*, 25(2), 105–120.
2. Intergovernmental Panel on Climate Change (IPCC). (2023). *AR6 Synthesis Report: Climate Change 2023*. https://www.ipcc.ch/report/sixth-assessment-synthesis-report/
3. Swiss Re Institute. (2024). *Global Catastrophe Losses: 2023 Report.* https://www.swissre.com/
4. United Nations Office for Disaster Risk Reduction (UNDRR). (2024). *Global Assessment Report on Disaster Risk Reduction.* https://www.undrr.org/
5. World Meteorological Organization. (2024). *State of Global Climate Report 2024.* https://public.wmo.int/

Chapter Four
Storm Intensification

António Guterres, UN SecretaryGeneral:

"Extreme storms ... will be the new normal of a warming world."

This stark warning—first issued in 2017 following Hurricane Irma—underscores that intensifying storms aren't anomalies but the expected outcome in our changing climate. Guterres emphasized that we already possess the tools, technology, and wealth to confront this—but what's missing is determination

A. Climate Dynamics

As global temperatures rise, so do the intensity, frequency, and unpredictability of storms. Climate change acts as an accelerant, warming oceans and saturating the atmosphere with moisture, which leads to more powerful hurricanes, cyclones, typhoons, and other extreme weather events. These intensified storms are no longer isolated anomalies—they are the emerging norm, threatening lives, infrastructure, economies, and ecological systems worldwide.

Storm intensification due to climate change is a growing and measurable phenomenon with far-reaching impacts worldwide. As the Earth's climate warms, the energy dynamics that govern storm behavior are shifting in ways that make hurricanes, typhoons, and other severe storms more powerful, longer lasting, and more destructive.

At the heart of this intensification is the warming of ocean waters. Tropical storms draw their energy from heat stored in the ocean's upper layers. As global sea surface temperatures rise, particularly in the tropical and subtropical zones, storms have more fuel to grow stronger. This additional thermal energy not only increases wind speeds within storm systems but also contributes to a trend known as rapid intensification. This means that storms can grow dramatically in power—sometimes escalating from a Category 1 to a Category 4 or 5 hurricane within just 24 hours.

Moreover, warmer oceans contribute to higher sea levels through both thermal expansion and melting ice sheets.

When storms form over elevated sea levels, the resulting storm surges are significantly higher, compounding the potential for coastal flooding and infrastructure damage.

The atmosphere's increased ability to hold moisture is another critical factor. For every degree Celsius that the global average temperature rises, the atmosphere can hold about 7 percent more water vapor. This surplus moisture translates into heavier rainfall during storm events, often resulting in flash floods, landslides, and infrastructure failures.

Climate change is also influencing the paths and intensities of storms. A documented poleward shift in tropical cyclone tracks has put new regions—previously considered outside the danger zone—at risk of being affected by landfall. Meanwhile, some storms are moving more slowly, which prolongs exposure to wind and rain for affected areas.

Hurricane Dorian in 2019 stalled over the Bahamas for more than 24 hours, causing catastrophic damage and loss of life. These changing behaviors are linked to altered jet stream patterns and weakening steering currents in the atmosphere.

There is also growing concern in the scientific community about the emergence of what could be termed "hypercanes"storms so intense that they would exceed the maximum threshold of today's Category 5 classification. While still theoretical, such scenarios are not impossible in a world that warms more than 2°C. Additionally, storms are forming earlier and later in the season and sometimes in regions not historically prone to such events, catching local governments and populations off guard.

Scientific consensus confirms the seriousness of these developments. The Intergovernmental Panel on Climate Change (IPCC), in its Sixth Assessment Synthesis Report released in 2023, noted with high confidence that the frequency of intense storms—particularly those classified as Category 3 to 5 systems—has increased. The report also confirmed that extreme rainfall associated with tropical cyclones is becoming increasingly common and that these trends are likely to intensify with continued warming. The World Meteorological Organization's 2024 global climate report corroborates these findings, highlighting the role of climate change in modifying storm behavior and amplifying risks to human settlements.

These intensifying storms are straining societies both economically and socially. Coastal cities are under increasing pressure to retrofit infrastructure, improve flood defenses, and revise zoning laws. The insurance industry has seen an escalation of claims from climate-intensified storms, with annual insured losses from major disasters exceeding $100 billion, according to the Swiss Re Institute's 2024 report. In many regions, the poorest and most vulnerable communities face disproportionate harm, underscoring the equity dimensions of storm resilience.

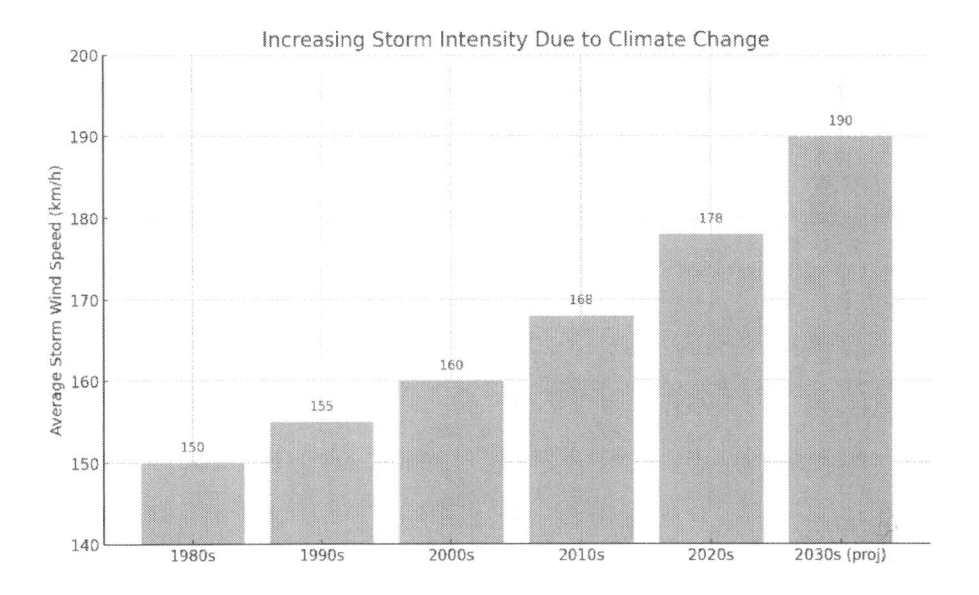

Increasing Storm Intensity Due to Climate Change

Decade	Average Storm Wind Speed (km/h)
1980s	150
1990s	155
2000s	160
2010s	168
2020s	178
2030s (proj)	190

B. What Does Storm Intensification Look Like?

Storm intensification, driven by climate change, is now a defining characteristic of extreme weather events worldwide. It is most clearly observed in the rapid escalation of hurricanes, typhoons, and cyclones, which grow more powerful, expansive, and destructive over a short span of time. This intensification manifests through a combination of physical traits—skyrocketing wind speeds, plummeting central pressure, massive rainfall, and escalating storm surges—all fueled by warmer ocean waters and a hotter atmosphere.

One of the most visible signs of intensification is the sharp increase in maximum sustained wind speeds. As storms intensify, their wind fields become increasingly violent, sometimes intensifying by dozens of miles per hour in a matter of hours. The Saffir-Simpson scale, which categorizes hurricanes from Category 1 to 5, is based on these wind speeds.

A storm that intensifies from a Category 1 hurricane with winds around 75 miles per hour to a Category 4 with winds over 130 miles per hour in just 24 to 36 hours can produce devastation far greater than initially expected. These rapid jumps in strength are increasingly common, leaving little time for vulnerable communities to prepare.

At the same time, meteorologists track a corresponding drop in minimum central pressure, the atmospheric pressure at the heart of the storm system. The lower the pressure, the more intense the storm. A sudden decrease of 10 to 20 millibars within a few hours signals that the storm is gaining strength quickly. For example, Hurricane Wilma in 2005 reached a record low pressure of 882 millibars in the Atlantic basin, a level of intensity made more likely by warming waters.

Another critical factor in storm intensification is sea surface temperature. Tropical storms typically require water temperatures of at least 26.5°C to form. Still, when ocean temperatures rise above 28 or 29 degrees Celsius—as is now frequently the case due to climate change, the potential for explosive storm growth increases dramatically. Scientists also look below the surface, measuring ocean heat content (OHC), which reflects the thermal energy stored deeper in the water column. Storms passing over high-OHC areas tend to retain or increase their strength even after churning up colder water, making these systems more resilient and potentially more dangerous.

The atmosphere itself also plays a crucial role. A warmer atmosphere can hold more moisture, leading to extreme precipitation during storm events. The rate of rainfall during intensified storms often exceeds six to eight inches in a single day, overwhelming urban drainage systems, flooding roads, and inundating homes. Satellite systems, such as NASA's Global Precipitation Measurement (GPM) network, enable scientists to track and quantify rainfall in real time, confirming that rainfall extremes are rising in tandem with global warming.

Another hallmark of intensified storms is their capacity to generate higher storm surges and stronger waves. These surges, caused by powerful winds pushing ocean water onto land, are measured with coastal tide gauges and buoys. As sea levels continue to rise due to the melting of glaciers and thermal expansion, storm surges reach farther inland, causing more destruction. In recent years, surges exceeding 10 to 15 feet have devastated coastal areas, particularly when combined with heavy rainfall and wind-driven waves.

The rate at which a storm intensifies is one of the most critical metrics for forecasters. When a storm increases its wind speed by 35 miles per hour or more within 24 hours, it is classified as undergoing rapid intensification. These events have become increasingly common. For instance, Hurricane Otis in 2023 intensified from a tropical storm to a Category 5 hurricane in less than 24 hours before striking Acapulco. Similarly, Hurricane Laura in 2020 and Hurricane Michael in 2018 both experienced rapid intensification shortly before landfall, leaving residents with little time to evacuate or prepare for the storm.

The scientific evidence confirms that this intensification trend is accelerating. Satellite data, ocean buoys, and hurricane reconnaissance flights all provide real-time metrics to monitor changes in storm strength. Wind speeds, central pressure, rainfall totals, storm surge heights, and ocean heat content are now routinely analyzed to assess the growing power of each storm. These measurements, when viewed in a historical context, show a clear and concerning pattern: climate change is not only increasing the frequency of extreme storms but making each one potentially more devastating than the last.

Understanding storm intensification through these indicators is crucial for enhancing forecasts, informing emergency responses, and guiding climate adaptation strategies. As the climate warms, the need to measure, anticipate, and prepare for rapidly intensifying storms has never been more urgent.

In sum, storm intensification is no longer a hypothetical risk—it is a present-day reality shaped by human-driven climate change. Its consequences are unfolding rapidly, calling for equally urgent action in adaptation, infrastructure resilience, equitable planning, and emissions reduction.

C. Examples of Storm Intensification

One of the most dramatic examples of this trend was Hurricane Harvey (2017), which brought catastrophic rainfall to Houston, Texas. Scientific analyses have since confirmed that climate change made Harvey's rainfall at least three times more likely and up to 38% more intense. The storm caused over $125 billion in damage, exposed weaknesses in urban flood control systems, and displaced tens of thousands of residents.

Recovery efforts revealed sharp disparities between low-income communities and wealthier neighborhoods, highlighting the intersection of climate vulnerability and social inequity. In response, Houston implemented climate resilience strategies, including the Bayou Greenways 2020 plan, which expanded green infrastructure for flood mitigation by restoring wetlands and converting bayous into public parklands that absorb and redirect stormwater.

On the other side of the globe, Cyclone Idai (2019) devastated Mozambique, Zimbabwe, and Malawi, killing over 1,300 people and affecting more than three million. The storm submerged entire towns and crippled food systems. In Mozambique, adaptation took the form of rebuilding housing on stilts, establishing early-warning radio systems in local dialects, and relocating entire communities away from floodplains.

Non-governmental organizations and the government collaborated to co-develop localized climate adaptation plans rooted in traditional knowledge and climate forecasting, showing how place-based responses can blend scientific and cultural approaches.

In the Pacific, intensified typhoons have repeatedly battered the Philippines, most notably Super Typhoon Haiyan (also known as Typhoon Yolanda) in 2013. With wind speeds exceeding 190 mph and storm surges reaching 20 feet, Haiyan killed over 6,000 people and displaced more than 4 million. The disaster led to one of the most significant climate displacement crises of the decade. In the aftermath, the Philippine government partnered with international organizations to establish climate resilient "No Build Zones," redesign school buildings as emergency shelters, and integrate climate risks into national urban planning codes. The country also launched Project NOAH (Nationwide Operational Assessment of Hazards), a real-time flood and typhoon tracking system that utilizes satellite data and geospatial modeling to inform evacuation and emergency response efforts.

In the United States, New York City's experience with Superstorm Sandy (2012) fundamentally changed its approach to storm adaptation. Sandy caused $70 billion in damage and highlighted how climate risks disproportionately affect public housing residents and critical infrastructure, such as subways and hospitals. In response, the city developed the "Resilient Neighborhoods" initiative and "Rebuild by Design," a planning competition that produced projects like "The Big U," a multi-billion-dollar storm surge barrier and park system encircling Lower Manhattan.

These projects embrace multifunctional infrastructures that are protective during storms and livable during calm weather. Sandy also led to reforms in federal disaster funding, enabling investments in long-term adaptation instead of short-term repairs.

In Bangladesh, where tropical cyclones regularly threaten the low-lying delta, adaptation has taken a different form. A massive network of cyclone shelters raised roads, and embankments has been developed over the past three decades. These interventions have dramatically reduced cyclone-related deaths, from hundreds of thousands of thousands in the 1970s to fewer than a hundred in more recent storms, such as Cyclone Amphan (2020). Bangladesh's Coastal Embankment Improvement Project combines engineered protections with nature-based solutions such as mangrove restoration, which not only buffers storm surges but also strengthens biodiversity and local livelihoods.

Storm intensification demands a shift in how societies perceive risk. It is no longer sufficient to prepare for "100-year storms" as rare events. The new climate reality requires adaptive governance that prioritizes vulnerable populations, integrates climate projections into building and zoning codes, and uses nature-based infrastructure alongside advanced technology. Artificial intelligence is beginning to play a role in predictive modeling and emergency logistics, but it must be paired with community-driven planning and culturally grounded resilience strategies.

The global response to intensified storms underscores a critical truth: adaptation is not merely a technical challenge but a profound social and political one. Those most at risk are those with the least resources, and equitable adaptation will require systemic reforms, climate justice frameworks, and international solidarity. As the climate continues to destabilize historical storm patterns, the imperative to adapt intelligently, inclusively, and urgently becomes more pressing than ever.

D. AI Examples Applied to Storm Intensification: How Artificial Intelligence Is Shaping Climate Risk Awareness.

Artificial intelligence is revolutionizing how we visualize and understand storm intensification in a warming world.

By integrating massive datasets—such as satellite imagery, atmospheric readings, ocean temperatures, and historical cyclone tracks—AI tools generate highly detailed visual models that simulate the evolution, severity, and impacts of extreme storms in real time or into the future. These visualizations are being utilized by planners, meteorologists, emergency responders, and the public to prepare for and adapt to increasingly intense storm events.

Here are key types of AI-generated visualizations currently in use, with examples:

E. Storm Track and Intensity Projection Maps

Using machine learning trained on decades of storm behavior, these models create predictive paths for tropical cyclones and estimate wind intensity and rainfall distribution hours or even days in advance. Unlike traditional models, AI-enhanced projections adjust dynamically as new sensor data (e.g., satellite and buoy readings) stream in.

Example: IBM's The Weather Company combines AI with radar and satellite data to generate real-time animated maps that show potential paths of hurricanes, including probability bands for landfall and precipitation volumes. Municipalities along the Gulf Coast have used these maps to guide evacuation zones before storms.

2. Sea Level Rise and Surge Inundation Models

AI-enhanced visualizations simulate how intensified storms interact with rising sea levels, resulting in dynamic flood maps that can be layered over urban infrastructure, transportation corridors, and vulnerable communities.

Example: Google's AI Flood Hub uses deep learning to map storm surge scenarios in high-risk coastal regions such as Bangladesh, Florida, and Indonesia. By integrating rainfall, terrain, soil saturation, and tide cycles, it generates minute-by-minute flood depth animations that local planners can use to pre-position aid and identify relocation zones.

3. Urban Stormwater Simulation Models

AI models, such as DeepRain or BlueDot Resilience, utilize 3D urban models and precipitation forecasts to simulate the movement of stormwater through city streets, storm drains, and low-lying neighborhoods.

These simulations often include animated flyovers that show how floodwaters rise block by block, helping to prioritize infrastructure investments.
Example: In Tokyo, city planners use AI-enabled 3D flood visualization platforms to simulate simultaneous riverine and pluvial flooding under typhoon conditions. The city's flood-control tunnels and reservoirs are designed and updated based on these models.

4. Climate Projection Animation Tools

These visualizations utilize IPCC and CMIP6 climate models to render storm intensification over decades, facilitating the visualization of long-term risks. AI enhances these projections by reducing computational error, interpolating missing data, and producing region-specific outputs.
Example: Climate Central's Picturing Our Future tool, now enhanced with AI, provides side-by-side visual comparisons of coastal cities (e.g., Miami, Lagos, Kolkata) under different warming scenarios (1.5°C vs. 3°C). Users can explore what storm surges would look like in these futures, with overlays showing schools, hospitals, and critical infrastructure at risk.

5. Thermal Ocean Mapping and Cyclogenesis Prediction

AI-enhanced ocean heat content maps help predict where storms are likely to form and intensify. These tools utilize satellite-derived sea surface temperature data, combined with subsurface ocean profiling, to visualize "hot spots" for potential cyclogenesis.
Example: The European Space Agency (ESA) has developed deep-learning models that predict cyclone formation up to two weeks in advance, rendering 4D visualizations that show how storm strength correlates with subsurface warming. These are used in early warnings in the Caribbean and Western Pacific.

6. Storytelling Visuals for Public Education

AI-generated visuals are also being used to translate data into emotionally resonant, interactive storytelling formats. These include immersive VR/AR environments, dynamic dashboards, and personalized flood risk maps.
Example: The First Street Foundation's RiskFactor™ AI-driven platform allows users in the U.S. to enter their address and receive a personalized, animated storm and flood risk report for the next 30 years. These visualizations have influenced homeowner decisions, insurance markets, and local planning ordinances.

AI-powered visualizations of storm intensification are not just technological marvels they are vital decision-making tools. They allow communities to see what's coming, how it might unfold, and what can be done to prepare. From predictive evacuation to climate-resilient infrastructure design, these tools are transforming abstract climate data into actionable insight, especially when embedded into planning platforms and community education campaigns. In the age of climate disruption, seeing is not only believing—it's surviving.

F. Conclusion

Storm intensification demands a paradigm shift in planning. It is no longer about returning to the pre-disaster status quo but about designing for uncertainty, equity, and transformation. This toolkit is not a one-size-fits-all solution but a flexible framework that must be customized to the unique geography, culture, and history of each place. As frontline responders to the climate crisis, planners hold the keys to resilience—and the responsibility to wield them justly and boldly.

Advocacy Brief

Addressing Storm Intensification in the Era of Climate Change

Prepared for policymakers, urban planners, educators, emergency responders, and community leaders

Date: June 2025
Issue Overview

Storm intensification is one of the most visible and destructive manifestations of climate change. Driven by rising global temperatures, warmer ocean surfaces, and atmospheric instability, storms—particularly hurricanes, typhoons, and tropical cyclones—are becoming more powerful, more frequent, and more unpredictable. This intensification has devastating impacts on human lives, infrastructure, economies, and ecosystems, especially in coastal and low-income communities.

The Intergovernmental Panel on Climate Change (IPCC, 2023) confirms that the frequency of Category 4 and 5 storms has increased since the 1980s and that extreme rainfall events associated with these storms are intensifying globally. Without aggressive adaptation and mitigation, storm-related losses will continue to escalate.

Key Drivers of Storm Intensification

Warming Oceans

- *Rising sea surface temperatures provide more energy for storm formation and escalation. This has led to a higher number of rapidly intensifying storms that increase in strength by at least 35 mph within 24 hours—leaving communities with little time to prepare.*

Higher Atmospheric Moisture

- *A warmer atmosphere holds more water vapor, resulting in extreme precipitation. The enhanced rainfall increases flood risks even in areas far from coastlines, threatening water systems, roads, and housing.*

Rising Sea Levels and Storm Surge

- *Sea level rises amplify the height and reach of storm surges. Combined with intense rainfall, this creates compound flooding events that overwhelm infrastructure and displace entire populations.*

Slower-Moving Storms and Altered Paths

- *Changing jet stream dynamics have caused storms to stall over vulnerable regions, intensifying damage and delaying recovery. Some storms are shifting poleward or appearing outside traditional cyclone zones.*

Urgent Needs and Equity Imperatives

- *Storm intensification does not impact all communities equally. Marginalized and low-income populations often reside in the most flood-prone areas and lack the resources for recovery. The elderly, disabled, and children are especially at risk. Climate adaptation policies must prioritize equity, access to early warning systems, and local capacity building.*

Policy and Action Recommendations

Advance Climate-Resilient Infrastructure

- *Invest in flood-resistant public infrastructure, including elevated roads, stormwater systems, seawalls, and green infrastructure. Prioritize nature-based solutions such as wetland restoration and mangrove reforestation.*

Update and Enforce Land Use Planning

- *Revise zoning laws to prohibit new development in high-risk flood zones. Provide incentives for relocation and elevation of vulnerable structures. Integrate storm intensification projections into comprehensive plans and building codes to enhance resilience.*

Fund Early Warning and Forecasting Technologies

- *Expand the use of AI-powered prediction systems, satellite monitoring, and real-time alert platforms. These systems can enhance preparedness and response, particularly in underserved regions.*

Protect Frontline Communities

- *Ensure that adaptation funds are equitably distributed to historically excluded groups. Support community-led resilience hubs, emergency shelters, and access to affordable insurance coverage.*

Integrate Climate Literacy and Public Education

- *Launch campaigns and school curricula that increase awareness of climate-related storm risks. Educated citizens are more likely to take preparedness measures, vote for climate action, and reduce vulnerability.*

Strengthen Federal and International Coordination

- *Collaborate across federal, state, tribal, and international lines to share climate data, response strategies, and funding. Advocate for the inclusion of storm-intensification measures in global climate treaties and development goals.*

Supporting Evidence

- *IPCC (2023). AR6 Synthesis Report. High confidence in increased frequency and intensity of tropical cyclones due to anthropogenic warming.*
- *World Meteorological Organization (2024). State of Global Climate Report. Highlights that over 80% of major global disasters in 2023 were water- and storm-related.*
- *Swiss Re Institute (2024). Reports over $115 billion in insured losses from climate-amplified storms in 2023 alone.*
- *NOAA and NASA (2025). New models confirm a 40% increase in rapidly intensifying storms over the past two decades in the Atlantic basin.*

Call to Action

Storm intensification is not a distant threat; it is a current crisis reshaping our coasts, cities, and communities. Delaying action will lead to exponentially higher recovery costs, human displacement, and environmental degradation. We urge government leaders, urban planners, emergency managers, and educators to adopt forward-looking, equity-centered strategies that strengthen community resilience in the face of increasingly severe storms.

The time to act is now—before the next storm makes landfall.

Toolkit for Planners

As storms intensify due to climate change, planners must lead the development of forward-looking, equitable, and ecologically sound adaptation strategies. This toolkit equips municipal, regional, and land-use planners with the foundational components for building climate-resilient communities in the face of more frequent and destructive storms.

1. Climate Risk and Vulnerability Assessment

The first step in any adaptation strategy is to conduct a localized climate risk assessment. Use downscaled climate models, hydrological forecasts, and historical storm records to identify areas at high risk of flooding, storm surge, landslides, and infrastructure failure. Combine this with a vulnerability assessment that maps social indicators—such as income, race, language, age, and housing type—to understand who will be most affected. Tools like FEMA's National Risk Index, NOAA's Sea Level Rise Viewer, and the IPCC's AR6 Urban Atlas can guide this step.

2. Integrated Land-Use Planning and Zoning Reform

Storm-intensified flooding and erosion call for land-use regulations that prevent new development in high-risk zones. Planners should revise zoning codes to:

- Establish "No Build" zones in floodplains, coastal wetlands, and erosion-prone slopes.
- Encourage cluster development and transfer of development rights (TDR) to redirect growth to safer zones.
- Incentivize green infrastructure through overlays and zoning bonuses.
- Implement adaptive zoning that allows for strategic retreat and the establishment of habitat migration corridors.

3. Nature-Based Solutions and Green Infrastructure

Integrate ecological resilience into planning by restoring and protecting natural buffers. These include:

- Wetlands and mangroves for storm surge absorption
- Urban forests and bioswales to absorb rainwater
- Green roofs and permeable surfaces for stormwater retention
- River corridor rewilding to create floodable open space

Cities like Rotterdam, New York, and Singapore have successfully blended these into urban design, improving resilience while enhancing livability.

4. Building Code Modernization and Infrastructure Resilience

Update building codes to reflect new storm realities:

- Mandate elevated foundations in flood-prone zones
- Require wind-resistant materials in cyclone and hurricane zones
- Install backflow valves, sump pumps, and waterproofing for critical infrastructure
- Prioritize microgrid-capable energy systems to ensure continuity of power during outages

Pair code reform with public infrastructure upgrades, such as hardened substations, raised transit hubs, and floodproofed hospitals.

5. Emergency Preparedness and Early Warning Systems

Planners should collaborate with emergency managers to integrate:

- *Smart sensor networks and AI-powered early warning systems for rainfall, wind, and flood levels*
- *Resilient communications infrastructure for multilingual alerts*
- *Evacuation route planning embedded in urban design*
- *Real-time GIS dashboards for emergency coordination*

Invest in community-based drills and preparedness education, especially in neighborhoods with limited mobility or digital access.

6. Equitable Relocation and Managed Retreat Frameworks

Where physical risk cannot be mitigated, planners must facilitate just and voluntary retreat:

- *Identify communities facing chronic inundation and prioritize community-led relocation planning*
- *Offer legal, financial, and psychological support for displaced residents*
- *Create "receiving areas" with affordable housing, services, and access to jobs*
- *Establish public land banks to acquire and restore vacated lands as green infrastructure*

Programs like New York's "NY Rising" and Louisiana's "LA SAFE" offer tested models.

7. Cross-sectoral and Intergovernmental Coordination

Storm adaptation requires cooperation across multiple scales:

- *Aligning regional transportation, housing, and environmental agencies under shared resilience goals*
- *Create inter-municipal stormwater utilities to manage watersheds holistically*
- *Coordinate with tribal governments and Indigenous communities, respecting sovereign planning authority*
- *Partner with state and federal agencies to leverage grants, such as FEMA BRIC, HUD CDBG-DR, and EPA resilience funding*

8. Financing and Implementation Strategies

Adaptation costs can be high, but inaction is far more expensive. Planners should explore:

- *Climate resilience bonds and municipal green bonds*
- *Public-private partnerships for resilient infrastructure*
- *Insurance incentives for resilient retrofits*
- *Development impact fees to fund stormwater management*

Include climate adaptation criteria in capital improvement plans (CIPs) and budget frameworks to ensure effective planning and implementation.

9. Community Engagement and Participatory Planning

Adaptation must be inclusive. Use community mapping, participatory budgeting, and visioning workshops to co-design solutions. Support youth, elders, renters, and marginalized communities with stipends, interpretation services, and non-technical educational materials. Trust and transparency will determine success.

10. Monitoring, Evaluation, and Adaptive Management

Establish metrics for success—flooded acres reduced, lives saved, infrastructure downtime avoided—and use them to evaluate projects. Incorporate adaptive management by revisiting plans after each major storm to assess performance and improve future interventions. Build a culture of continuous learning through data collection, academic partnerships, and community storytelling.

Resources

1. *Intergovernmental Panel on Climate Change (IPCC). (2023). Sixth Assessment Synthesis Report. https://www.ipcc.ch/report/sixth-assessment-synthesis-report/*
2. *World Meteorological Organization (WMO). (2024). State of the Global Climate 2024. https://public.wmo.int/*
3. *Swiss Re Institute. (2024). Global Catastrophe Losses 2023 Report. https://www.swissre.com/*
4. *NOAA National Hurricane Center. (2024). Tropical Cyclone Reports. https://www.nhc.noaa.gov/*

5. Knutson, T. R., Camargo, S. J., Chan, J. C. L., Emanuel, K. A., Ho, C. H., Kossin, J. P., ... & Sugi, M. (2023). *Tropical Cyclones and Climate Change Assessment: 2023 Update. Bulletin of the American Meteorological Society, 104(3)*, E567–E589. https://doi.org/10.1175/BAMS-D-22-0133.1
6. Patricola, C. M., & Wehner, M. F. (2023). *Anthropogenic Influences on Major Tropical Cyclone Rainfall. Nature Communications, 14(1)*, 1421. https://doi.org/10.1038/s41467-023-37159-8
7. Emanuel, K. A. (2023). *Climate and the Behavior of Tropical Cyclones. Annual Review of Earth and Planetary Sciences, 51*, 91–117. https://doi.org/10.1146/annurev-earth-032822-051949
8. NASA Earth Observatory. (2024). *Hurricane Intensification and Ocean Heat Content.* https://earthobservatory.nasa.gov/features/OceanHeat
9. Zhang, Z., Wang, Y., & Mei, W. (2024). *Increasing trend in rapid intensification of tropical cyclones in a warming world. Geophysical Research Letters, 51(2)*, e2023GL102412. https://doi.org/10.1029/2023GL102412

Chapter Five

Drought

Ibrahim Thiaw, Executive Secretary of the UN Convention to Combat Desertification (UNCCD)

"Droughts are not just natural disasters. They are the result of poor land and water management. Climate change is intensifying them, but we still have a choice: restore land and adapt."

A. Climate Context and Urgency

Drought is intensifying rapidly across diverse regions due to climate change, disrupted precipitation cycles, and widespread land mismanagement. The World Meteorological Organization (2024) reports that over 3 billion people now live under water stress, and projections suggest that as many as 700 million individuals could be displaced by drought-related conditions by 2030. Once considered stable, regions such as the western United States, Mediterranean Europe, Sub-Saharan Africa, and Southeast Asia are now experiencing critical water shortages. The IPCC Sixth Assessment Report (2023–2025) emphasizes that escalating drought events are symptoms of both rising atmospheric temperatures and declining ecosystem resilience.

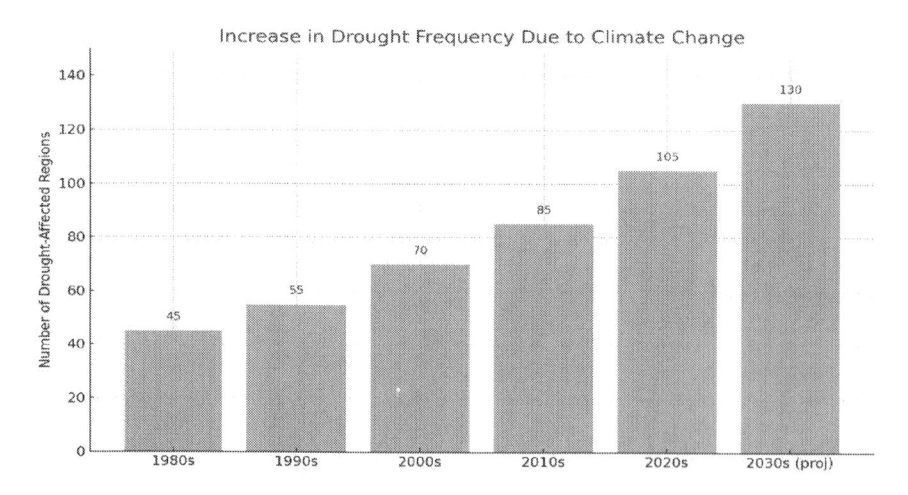

Increase in Drought Frequency Due to Climate Change

Decade	Number of Drought-Affected Regions
1980s	45
1990s	55
2000s	70
2010s	85
2020s	105
2030s (proj)	130

B. Objectives of a Drought Adaptation Strategy

This blueprint for drought adaptation aims to ensure sustainable water access, strengthen ecological systems that naturally regulate hydrological cycles, and establish cross-sectoral coordination among energy, agriculture, public health, and land-use governance. The approach is grounded in the recognition that addressing drought requires integrated systems thinking and long-term planning.

C. Frameworks for Implementation of Drought Adaptation Strategies

Drought adaptation begins with the transformation of land use practices. Bioregional land use planning must guide decisions to protect aquifers, recharge zones, and watershed areas, thereby ensuring the sustainable management of these critical resources.

Development should be limited or redirected in zones already experiencing groundwater depletion, salinization, or Desertification. Municipal design standards can incorporate green infrastructure, including bioswales, permeable pavements, rain gardens, and xeriscaping—landscaping methods that use native, drought-resistant vegetation to reduce water demand and improve retention.

Water governance requires a paradigm shift that moves away from private commodification toward shared stewardship. Municipalities can implement drought ordinances to regulate consumption during periods of crisis and establish updated water rights regimes that reflect both the needs of ecosystems and current climate realities. Policies should ensure the restoration of environmental flows in rivers and wetlands, as these flows are critical to maintaining biodiversity and long-term water availability.

Public health systems must be adapted to respond to drought-related risks, including dehydration, respiratory illness from dust exposure, and increased rates of malnutrition. Proactive planning can include early warning systems, public cooling stations, and strengthened support for medical systems during extreme heat events. Food systems should transition toward drought-resilient agricultural practices, including dryland farming, regenerative soil management, and the preservation of heirloom seeds. Promoting local crops and diversified diets can reduce dependence on water-intensive production systems.

Educational systems play a critical role in preparing the next generation for water-scarce futures. Schools and universities can integrate drought-focused curricula that combine climate science, hydrology, geography, and civic planning to address this issue. Experiential learning methods—such as community mapping, groundwater testing, and school gardens—can engage students in real-world adaptation efforts. Community workshops and citizen science programs can complement formal education, enabling residents to monitor soil conditions, assess rainfall patterns, and document changes in water quality.

Technological innovation must support adaptive capacity. Climate-smart irrigation systems that reduce runoff and increase efficiency, combined with AI-driven monitoring of water tables and soil moisture, can inform data-driven policy decisions. Municipalities can deploy leak detection systems to prevent infrastructure losses and mandate rooftop rainwater harvesting and greywater reuse in residential and commercial buildings.

Legal and Policy Instruments

Effective drought governance requires enforceable legal mechanisms. Municipalities can mandate drought vulnerability assessments for all new development projects and integrate climate-health co-benefit evaluations into environmental reviews. Legislation can require both Personal Environmental Impact Statements (PEIS) and community-level Drought Adaptation Plans to ensure that individuals and planners consider the long-term consequences of water-related decisions. Development permits should be linked to demonstrated resilience measures, including water budgeting, soil retention plans, and ecosystem restoration strategies.

Funding sources must be aligned with the scale of the challenge. Climate resilience bonds, public adaptation grants, and municipal financing tools can be designed to prioritize long-term infrastructure and planning investments. Tax codes can be amended to support conservation retrofits, aquifer recharge systems, and drought-resilient farming practices. Community planning initiatives should receive sustained support through technical assistance, capacity building, and partnerships with universities and nonprofit institutions.

Monitoring Progress and Outcomes

Monitoring systems must track indicators such as reductions in per capita water use, the percentage of landscapes converted to drought-adapted vegetation, and public access to potable water during dry periods. Municipal dashboards and mobile reporting tools can enhance transparency, enabling residents to track water use patterns and report drought conditions. These systems can support evidence-based policy corrections and build a culture of shared responsibility and adaptive learning.

International and Local Models

Recent examples illustrate effective responses to worsening drought. In Santa Fe, New Mexico, climate adaptation is integrated into land use codes, and water budgeting is mandatory for all new construction projects. Cape Town, South Africa, which faced the threat of a complete municipal water shutdown in 2018, has since restructured its water management systems through leak detection, conservation campaigns, and private-public infrastructure upgrades. In Chiapas, Mexico, indigenous communities have revitalized watershed health through the application of traditional knowledge, forest stewardship, and small-scale irrigation redesigns.

Drought reveals more than ecological strain; it exposes failures in governance, planning, and social foresight. This blueprint offers a regenerative path forward, built on science, legal tools, and human foresight. The goal is not only to survive conditions of water scarcity but to design systems capable of adapting, evolving, and regenerating in the face of disruption. In the Anthropocene, drought adaptation is not a choice. It is a collective responsibility.

D. Continental-Scale Groundwater Flow: New Knowledge

Recent simulations led by Princeton and the University of Arizona revealed that groundwater can travel up to 100 miles beneath the surface, with sub-surface residence times ranging from 10 to 100,000 years, before surfacing far from where recharge occurred. Similarly, studies show that deep groundwater from consolidated sediment layers (10–100 m deep) constitutes more than half of the baseflow in over half of the continental sub-basins—an insight that challenges earlier watershed-focused water-balance models. These findings demonstrate that regional drought resilience must account for deep aquifer transit, not just surface and near-river flows.

Hydrologic Connectivity Across Landscapes

A recently published review defines hydrologic connectivity as the movement of water, sediments, nutrients, and biota between landscape patches. Emerging strategies highlight the use of integrated drones, remote sensing, and ecological-hydrologic models to better map and manage multi-scale connectivity—from headwater catchments to inter-basin exchange. At the continental level, applying information-theory metrics, such as transfer entropy, helps quantify how hydrological fluxes propagate across large networks.

Surface–Subsurface Coupling at Scale
In China, the CONCN modeling platform now fully integrates surface water and groundwater at 30-arcsec (~1 km) resolution across the entire continent. A similar effort is underway in North America, where surface–subsurface coupling using ParFlow-like frameworks aims for high-fidelity hydrogeologic predictions essential for drought assessment and aquifer recharge planning.

Groundbreaking platforms like HydroTrace now leverage AI and attention mechanisms to model streamflow at continental scales with up to 98% accuracy (Nash–Sutcliffe Efficiency).

They can also interpret spatial-temporal drivers, such as glacier melt or monsoonal flow, and are accessible via LLM interfaces at arxiv.org. Physics-aware ML frameworks further integrate first-principles physics with data-driven learning to enhance model interpretability and transferability, as described at arxiv.org.

Land-Atmosphere Feedback and Soil Moisture Depletion
Continental coupling isn't just subsurface. Lateral terrestrial water flows (e.g., evapotranspiration and soil moisture) have been shown to increase precipitation by ~3–6% in Europe and West Africa, influencing land-to-atmosphere feedback.

Meanwhile, global studies indicate that Earth has lost over 2,000 gigatons of terrestrial water storage in the past 20 years, including soil, rivers, and lakes, which has caused increases in drought frequency, agricultural stress, sea-level rise, and even shifts in Earth's axis of rotation. These changes appear irreversible within human lifespans unless water use and climate heating are significantly addressed.

E. Integrating AI and Groundwater Pathway Mapping in Drought Adaptation

1. AI Modeling in Hydrological Forecasting and Drought Prediction

Artificial Intelligence is revolutionizing the way we understand, predict, and manage hydrological systems. Machine learning (ML) models—profound learning frameworks—are now used to simulate river discharge, evapotranspiration, aquifer recharge, and soil moisture across entire continents, even under data-sparse conditions.

Key Developments:

HydroTrace (2024), an open-access AI platform, models streamflow at a continental scale using transformer-based architectures (similar to those used in language models). It integrates satellite imagery, meteorological inputs, topography, and land cover to predict water flow with an accuracy of up to 98%, as measured by the Nash–Sutcliffe Efficiency coefficient.
Physics-informed neural networks (PINNs) integrate hydrological principles with the flexibility of machine learning. They encode conservation laws (e.g., mass balance, Darcy's Law) directly into the model architecture, ensuring that predictions remain physically plausible even in regions with limited data.

PINNs are increasingly applied in modeling aquifer recharge dynamics, wetland retention, and floodplain storage across large domains.

Probabilistic drought forecasting tools, such as those developed using Bayesian neural networks, now provide uncertainty quantification alongside predictions. This helps planners assess not only expected outcomes but also risk scenarios under multiple climate pathways.

AI is also being used for real-time drought monitoring. Integrated with ground sensors and satellite feeds (e.g., SMAP and GRACE), models can track soil moisture depletion, vegetation stress, and aquifer changes in near real-time. These models are critical for triggering early warning systems, emergency declarations, and water use restrictions.

2. Groundwater Pathway Mapping: Tracing the Invisible Water Highways

Groundwater does not respect political or watershed boundaries. Recent research has unveiled that deep aquifers can carry water laterally for tens to hundreds of miles, often over centuries or millennia. Mapping these invisible water highways is essential for resilient land use and water governance.

Key Innovations:

Recent studies by the University of Arizona and Princeton (2024) demonstrated that water infiltrating consolidated sediment layers at depths of 10–100 meters may surface more than 160 kilometers from its recharge point. This discovery upends traditional water basin management, which has long assumed localized flow paths.

Isotope hydrology now enables researchers to date groundwater and track its journey using isotopes such as tritium, deuterium, and carbon-14. These markers distinguish between modern and ancient groundwater, helping to determine whether an aquifer is actively recharging or in a state of fossil decline.

GRACE satellite data, which measures gravity anomalies from space, provides a powerful tool to assess changes in terrestrial water storage. Combined with digital elevation models and soil characteristics, this data allows for subsurface flow modeling at national and continental scales.

3D hydrogeologic modeling platforms such as MODFLOW 6, ParFlow, and China's CONCN are now run at kilometer-scale resolution across entire continents. These models incorporate lithology, topography, recharge rates, evapotranspiration, and human withdrawals to simulate groundwater movement in real-time scenarios.

Subsurface AI mapping tools now use convolutional neural networks (CNNs) trained on borehole data, well logs, and seismic imaging to predict underground water-bearing strata and preferential flow paths. These tools support municipal planners in locating new well sites, evaluating aquifer sustainability, and forecasting drawdown risks.

3. Applications for Drought Planning and Land Use

In practical terms, integrating AI modeling and groundwater pathway mapping into a municipal or regional drought plan allows for:

- Predictive allocation of water resources based on future drought probability and flow anomalies.
- Zoning laws account for groundwater inflow from distant regions, preventing overdraft and saltwater intrusion in vulnerable basins.
- Design of managed aquifer recharge (MAR) systems that take advantage of known recharge zones.
- Real-time adjustments to irrigation, industrial water use, and public water supply based on updated forecasts.
- Avoidance of costly infrastructure projects in zones where deep water movement indicates unsustainable sourcing.

Understanding and visualizing groundwater flows and surface interactions at scale is essential for 21st-century drought governance. The convergence of AI, satellite remote sensing, and subsurface modeling opens a new era of predictive hydrology—where adaptation decisions are not reactive but anticipatory. These tools help us treat water as a dynamic, interconnected part of the planetary system rather than a static, local commodity. Drought resilience will increasingly depend on how well we can model and manage these hidden flows.

AI VISUALIZATIONS

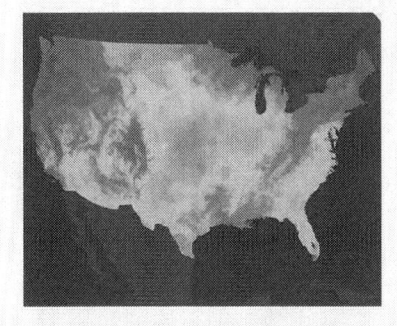

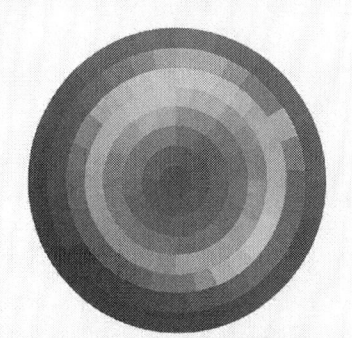

STREAMFLOW PREDICTION **DROUGHT FORECAST**

GROUNDWATER RECHARGE **STREAMFLOW ANOMALY**

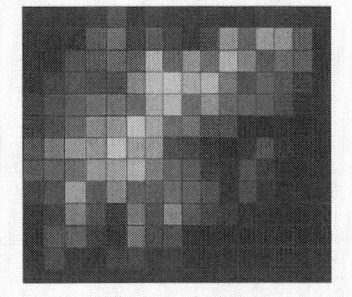

1. Streamflow Prediction (Top Left)

This heatmap displays projected river and stream discharge levels across the continental United States, generated using AI models such as HydroTrace or LSTM-based neural networks.

- Warmer colors (red and orange) indicate regions of high predicted streamflow.
- Cooler colors (blue and green) indicate low or dry streamflow conditions.
- These predictions are based on real-time meteorological inputs (precipitation, temperature), satellite data, and topographic features.

2. Drought Forecast (Top Right)

This circular plot represents a spatially resolved drought intensity forecast modeled using probabilistic or ensemble machine learning techniques.

- Concentric bands simulate severity zones, with the center (dark red) indicating the most severe drought projection.
- The model uses historical drought patterns, soil moisture levels, vegetation health indices (NDVI), and atmospheric anomalies to produce short- and long-term drought risk maps.

3. Groundwater Recharge (Bottom Left)

This heatmap illustrates modeled groundwater recharge rates, indicating the amount of water that percolates into aquifers after rainfall or snowmelt.

- High-recharge zones (red and yellow) are areas where aquifers are likely to be replenished.
- Low-recharge zones (blue/green) are at risk of long-term drawdown or over-extraction.
- These predictions combine land cover data, soil permeability, geology, evapotranspiration rates, and hydrologic history, often modeled with convolutional neural networks (CNNs) and integrated surface–subsurface simulations.

4. Streamflow Anomaly (Bottom Right)

This pixelated grid map shows anomalies in streamflow—differences between expected and observed values.

- Red or orange blocks signify regions with lower-than-expected flow, often indicating early signs of drought or hydrological stress.
- Blue/green blocks indicate above-normal streamflow, possibly due to snowmelt surges, upstream release, or heavy rainfall.
- Anomaly detection is key for early warning systems and is frequently modeled using Bayesian learning, ensemble forecasts, or recurrent neural networks.

These AI-generated visuals provide high-resolution, real-time insights for decision-makers, planners, and emergency managers, enabling anticipatory governance rather than reactive response.

F. Examples of Drought Management

Santa Fe, New Mexico — Basin-Wide Water Strategy with Subsurface Integration
Overview

Santa Fe and Santa Fe County collaborated with the Bureau of Reclamation and Sandia Labs on the Santa Fe Basin Study (2015, updated 2019). This basin-scale assessment evaluated climate impacts, surface and groundwater interactions, and longterm supply/demand under projected drought scenarios

AI & Modeling Applications
Using hydrologic simulation tools, they modeled future streamflow and aquifer conditions, assessing water deficits tied to both surface sources (Santa Fe River and imported water) and groundwater from local wellfields.
Groundwater Pathway Mapping

As part of their watershed resilience plan, Santa Fe invested in Managed Aquifer Recharge (MAR), directing stormwater into recharge zones identified via subsurface mapping and infiltration studies—linking green infrastructure (bioswales, permeable surfaces) with deeper aquifer replenishment.

Outcomes & Tools

- Aquifer Storage and Recovery (ASR) pilot initiatives help buffer seasonal variability.
- Subsurface mapping, including 3D models and geologic data, guides recharge projects and identifies safe withdrawal zones.
- The Basin Study results inform zoning rules and drought ordinances aimed at maintaining minimum aquifer depths during extreme drought.

Cape Town, South Africa — "Day Zero" Avoidance through Groundwater Diversification and AIEnhanced Monitoring

The 2015–2018 drought nearly triggered "Day Zero." Cape Town responded with aggressive water restrictions, leak-fixing campaigns, and new supply sources. To complement depleted surface dams, the city identified strategic aquifers for augmentation. Detailed hydrogeologic assessments traced suitable deep groundwater sources, incorporating isotope analysis and well logs to evaluate recharge rates and extraction capacity. Recent academic studies used machine learning (Random Forest, SVM, LSTM) to model groundwater-level fluctuations in South Africa's West Coast aquifer system. These models predicted water table changes with median errors of less than 0.4 m, offering powerful insights for managed recharge systems.

In Cape Town, remote sensing combined with AI helps confirm recharge zones and pressure dynamics within newly tapped aquifers.

Practical Outputs

- Shallow and deep groundwater wells were activated as additional sources of supply.
- AI-driven detection systems, morning leak scans, and anomaly alerts supported efficient water use management.
- Integrated water balance models tied usage to supply reliability, enhancing demandmanagement policies.

Element Santa Fe Cape Town
Scale Basin & watershed-wide Urban/Rural hybrid, Models Physically based hydrologic & MAR ML models (RF, SVM, LSTM), remote sensing. Recharge Strategy Stormwater and bioswales using ASR Deep aquifers for supply augmentation. Tools used: 3D subsurface mapping, climate models, Isotope hydrology, AI groundwater prediction.

Why These Matter

1. Both cities demonstrate how mapping unseen groundwater pathways informs smart recharge and extraction, protecting aquifers from overdrafts.
2. AI-driven analysis enables dynamic prediction of groundwater levels and streamflow, empowering anticipatory drought response.
3. These methods embed subsurface hydrology into governance, ensuring legal and land-use planning incorporates deep aquifer knowledge.

G. Conclusion

These case studies demonstrate how integrating AI, groundwater science, and adaptive planning leads to resilient water systems in drought-prone contexts. They are waiting to be embraced by the climate adaptation generation.

Advocacy Brief

Securing Water Futures: A Drought Adaptation Imperative for Climate Resilience Executive Summary

As climate change intensifies, hydrological extremes are becoming longer, more frequent, and more destructive—threatening food security, public health, economic stability, and ecological integrity. From California to the Horn of Africa, millions face water scarcity driven by rising temperatures, altered precipitation, groundwater depletion, and land mismanagement. This brief calls for urgent and equitable adaptation measures that integrate climate science, Indigenous water knowledge, legal reforms, and technological innovation to safeguard communities, ecosystems, and future generations.

The Problem

Drought is no longer a temporary weather event—it is a structural crisis worsened by climate change, deforestation, aquifer over-drafting, and inefficient water use. According to the World Meteorological Organization (2024), the number of people affected by drought has increased by nearly 30% since 2000. The IPCC AR6 (2023) warns that severe droughts in semi-arid and Mediterranean regions are expected to become three to four times more frequent by 2050 unless adaptive governance measures are implemented.

Consequences include:

- *Food insecurity and crop failure (especially in vulnerable rainfed agriculture zones)*
- *Public health threats, such as heat-related illness and contaminated drinking water*
- *Conflict and displacement, especially in regions where water is politicized*
- *Irreversible ecosystem degradation, including drying wetlands and biodiversity collapse*

Drought disproportionately impacts low-income, Indigenous, and rural communities, who often lack access to climate-resilient infrastructure or political influence.

Strategic Recommendations for Adaptation

1. Legally Mandate Water Resilience Planning

Adopt municipal and regional ordinances requiring:

- *Personal and Organizational Water Impact Statements*
- *Enforceable Drought Management Plans in all development permits*
- *Recognition of water as a public trust resource not subject to unrestricted private control*

2. Invest in Nature-Based and Traditional Solutions

- *Restore wetlands, aquifers, and riparian zones to increase recharge and buffer heat*
- *Scale Indigenous and ancestral water stewardship, such as the Zuni waffle gardens or Andean canal systems*
- *Protect bioregions through localized, watershed-based planning*

3. *Advance Technological and Data-Driven Tools*

- *Support AI-enhanced groundwater monitoring and early warning systems*
- *Implement real-time drought dashboards integrating climate, soil moisture, and socioeconomic data*
- *Promote permeable urban design, rainwater harvesting, greywater reuse, and smart irrigation*

4. *Reform Water Rights and Allocation Systems*

- *Prioritize water justice by rebalancing outdated "first in time, first in right" doctrines*
- *Shift subsidies toward conservation-based agriculture and drought-resilient crops*
- *Enable community-based water governance boards, especially in underserved regions*

5. *Education, Public Health, and Civic Engagement*

- *Integrate drought awareness and resilience education into school and community curricula*
- *Fund public health outreach on heat illness, dehydration, and toxic exposure during droughts*
- *Support youth- and elder-led water justice organizing, particularly among frontline populations*

Policy Actions and Model Frameworks

- *California's Sustainable Groundwater Management Act (SGMA) offers a baseline for local groundwater planning.*
- *UN Convention to Combat Desertification (2024 Update) emphasizes land degradation neutrality and participatory governance.*
- *The AI-Drought Collaborative (2025) facilitates data sharing among public agencies, Indigenous nations, and universities to promote equitable drought resilience.*

Call to Action

Climate-driven drought is a slow-onset disaster demanding immediate, coordinated action. Governments at all levels must treat water security as the frontline of climate adaptation. That means reimagining infrastructure, reallocating rights, respecting Indigenous knowledge, enforcing resilience mandates, and empowering civic participation. Failure to act now will ensure a deeper ecological collapse and humanitarian suffering.

We urge elected officials, planners, educators, and funders to prioritize equitable drought adaptation—because without water, there is no resilience, no justice, and no future.

Planners Toolkit

I. Purpose of the Toolkit

This municipal toolkit provides a strategic framework for integrating cutting-edge hydrological science, AI-based forecasting, and land-use policy into drought adaptation. Designed for local governments, utilities, planners, and community leaders, the toolkit translates research and global case studies into actionable governance tools.

II. Core Principles

- *Treating water as a dynamic, interconnected system beyond jurisdictional boundaries.*
- *Plan for long-term water security under multiple climate and socioeconomic scenarios.*
- *Embed groundwater pathway mapping and AI modeling into development review processes.*
- *Prioritize public transparency, inter-agency collaboration, and place-based knowledge.*

III. Key Components

A. Groundwater Pathway Integration

- *Use geophysical and isotope hydrology data to map subsurface flows.*
- *Require groundwater source assessments for all significant new developments.*
- *Designate protected recharge zones in zoning and master plans.*
- *Establish buffer regulations around aquifer-contributing areas.*

B. AI-Driven Hydrological Forecasting

- *Incorporate AI tools, such as HydroTrace or LSTM-based models, for streamflow and recharge forecasting.*
- *Develop drought dashboards using remote sensing and ML soil moisture maps.*
- *Develop early warning systems utilizing real-time groundwater and evapotranspiration data.*

C. Legal and Policy Instruments

- *Mandate Drought Vulnerability Assessments in environmental reviews.*
- *Link building and land use permits to aquifer recharge potential and projected water availability.*
- *Enact Managed Aquifer Recharge (MAR) ordinances encouraging permeable infrastructure.*

D. Infrastructure and Green Design

- *Incentivize bioswales, infiltration basins, and permeable pavements through zoning and tax credits.*
- *Implement dual-pipe systems to reuse greywater in municipal and commercial buildings.*
- *Require rooftop rainwater harvesting in new developments over a specified size.*
- *in underserved neighborhoods.*

IV. Implementation Framework

Step 1: Mapping and Modeling

- *Use satellite, borehole, and hydrogeological data to establish a regional water balance model. Integrate deep recharge pathways, streamflow dynamics, and land use overlays.*

Step 2: Policy Alignment

- *Revise general plans, water master plans, and municipal codes to reflect the model's findings. Include buffer zones, aquifer protections, and bright zoning overlays.*

Step 3: Infrastructure and Capital Projects

- *Incorporate MAR systems into stormwater plans. Prioritize green infrastructure and establish funding pipelines through resilience bonds or federal adaptation grants.*

Step 4: Monitoring and Feedback

- *Develop digital dashboards and mobile tools for residents to report instances of low flow, contamination, and transitions between flood and drought conditions. Monitor impacts via AI-enhanced analytics.*

V. Sample Ordinance Language

"No development shall occur within a designated recharge zone without a hydrogeologic study demonstrating no net loss of aquifer recharge capacity. Permits shall be conditioned on the implementation of site-level MAR infrastructure."

VI. Contact and Technical Assistance

For implementation support, please get in touch with your regional water management district, a university water institute, or a national climate adaptation center.

Prepared by:
[Your Organization/Author Name]
Date: [Insert Date]
For more information, visit: [Insert URL or contact info]

Resources

1. IPCC. (2023). *AR6 Synthesis Report: Climate Change 2023.* https://www.ipcc.ch/report/sixth-assessment-synthesis-report
2. World Meteorological Organization. (2024). *State of Global Climate Report 2024.* https://public.wmo.int
3. United Nations. (2024). *Convention to Combat Desertification (UNCCD).* https://www.unccd.int
4. California Department of Water Resources. (2024). *Sustainable Groundwater Management Implementation Report.*
5. International Bioregional AI Partnership. (2025). *AI Tools for Water Security and Drought Resilience.* Aerts, J. C. J. H., & Botzen, W. (2025). Adaptation to Drought in Urbanizing Regions: Risk Governance and Planning Tools. *Climate Policy Journal, 25(1)*, 85–110.
6. U.S. Drought Monitor & NOAA NIDIS. (2024). *Drought Conditions and Impacts Summary.* https://www.drought.gov
7. UNESCO & IHP. (2023). *Water Security and Climate Resilience: Adaptation in Arid Zones.* Paris: United Nations Educational, Scientific and Cultural Organization.
8. Natural Resources Canada. (2024). *AI-Guided Drought Risk Mapping in Prairie Regions: Pilot Results.* Government of Canada.

9. *International Bioregional AI Partnership. (2025). AI and Climate Resilience: Drought Modules for Shared Watersheds. Global Climate Response Network.*
10. *Food and Agriculture Organization (FAO). (2024). Digital Agriculture for Drought Resilience: AI, Satellites, and Soil Sensors.* <u>https://www.fao.org</u>
11. *United Nations Development Programme (UNDP). (2024). Water Justice and Climate Resilience: A Rights-Based Framework.* <u>https://www.undp.org</u>
12. *Redsteer, M. H., & Jandreau, C. (2024). Indigenous Water Knowledge in Drought Adaptation: Lessons from the Southwest and Great Plains. Journal of Climate and Culture, 13(2), 49–70.*
13. *Ostrom, E., revisited in Clark, D. & Ibarra, L. (2025). Governing Water Commons Under Climate Stress: Local Strategies from Latin America and Africa. Water Policy & Society,*

Chapter six
WILDFIRES

Greta Thunberg, Climate Activist

"The world is literally on fire, and still leaders hesitate. Wildfires are no longer natural disasters—they are climate disasters. Adaptation without justice is just another form of denial."
— *Twitter, 2021*

As wildfires intensify due to prolonged droughts, extreme heat, and changing precipitation patterns—all of which are fueled by climate change—planners face mounting pressure to develop proactive adaptation strategies that protect lives, ecosystems, and infrastructure. Wildfire adaptation necessitates a shift from reactive suppression to systemic, long-term resilience rooted in land-use planning, ecological restoration, and community risk reduction.

A. What Is a wildfire?

A wildfire is an uncontrolled fire that rapidly spreads through natural vegetation such as forests, grasslands, or shrublands. These fires originate from both natural sources, such as lightning strikes, and human activities, including unattended campfires, downed power lines, or deliberate acts of arson. While some fires can play a natural role in ecological renewal, climate change has dramatically increased their frequency, intensity, and destructiveness. Rising temperatures, extended droughts, earlier snowmelt, and drier vegetation all contribute to creating ideal conditions for wildfire ignition and spread. In many regions, especially in the western United States, Mediterranean Europe, and parts of Australia and Canada, the wildfire season has expanded significantly, now lasting longer and becoming more challenging to manage.

Wildfires are measured using several key criteria, each providing insight into a different aspect of the fire's behavior or impact. One of the most common indicators is the burned area, typically reported in acres or hectares. This reflects the total land surface scorched or destroyed by the fire and is a critical metric for assessing its scale. For instance, the August Complex Fire in California, which burned in 2020, covered over one million acres, making it the largest in state history.

Another critical factor is fire intensity, which measures the energy output of the fire and is usually quantified in kilowatts per meter. This tells scientists and emergency responders how much heat the fire is generating and whether it's hot enough to cause crown fires—those that leap into treetops—or throw embers miles ahead of the fire line. Closely related is fire severity, which refers not just to the heat but to the degree of ecological damage. Fire severity measures the extent to which vegetation has been consumed and the depth to which the heat has affected the soil layers and root systems. It is often determined after the fire has passed, using satellite imagery or ground-based surveys to compare pre-and post-fire vegetation health through indicators such as the Normalized Burn Ratio (NBR).

Another dynamic measure is the rate of spread, which calculates how quickly the fire front is advancing across the landscape, typically expressed in kilometers or miles per hour. This rate depends on several interacting conditions: the type and dryness of the vegetation (called "fuel"), the slope of the land, and, especially, the wind speed. Fires traveling uphill or pushed by strong winds can move at alarming speeds, overwhelming firefighting efforts.

An operational measure often used by emergency agencies is the containment percentage, which indicates the rate of the fire's perimeter that has been successfully encircled and controlled through firebreaks or suppression activities. A fire that is 75% contained means that firefighters have established barriers or suppression lines around three-quarters of the fire's edge, though flames may still be burning within or near the uncontained portion.

Forecasting and assessing wildfire risks also rely on environmental models, such as the Fire Weather Index (FWI), a system initially developed in Canada and now used globally to predict wildfire potential. The FWI considers variables such as wind speed, temperature, humidity, and fuel moisture content to provide daily fire danger ratings. This tool helps governments issue public warnings and allocate firefighting resources in anticipation of extreme fire behavior.

In addition to these fire-focused metrics, wildfires are increasingly evaluated based on their public health and climate impacts, primarily through the measurement of airborne particulate matter such as $PM_{2.5}$ and PM_{10}. These fine particles, released in massive quantities during wildfires, can travel thousands of miles and pose significant risks to respiratory and cardiovascular health. Smoke from large wildfires also contributes significantly to global greenhouse gas emissions, releasing carbon dioxide, methane, and black carbon into the atmosphere, thereby exacerbating the climate crisis.

Modern wildfire monitoring combines satellite technologies, drones, and ground-based sensors. Satellites such as NASA's MODIS and VIIRS systems provide thermal imaging and detect hotspots across the globe, enabling real-time tracking of fire outbreaks. Drone technology enhances this capacity by offering high-resolution, low-altitude imaging and infrared data to map fire progression. On the ground, specialized sensors track changes in heat, moisture, and air quality. Increasingly, artificial intelligence is used to analyze these data streams, simulate fire behavior, and provide predictive models that can guide decision-making during emergencies.

In summary, wildfires are complex natural events made increasingly dangerous by anthropogenic climate change. Their measurement requires a combination of physical, ecological, and technological assessments. Understanding how they are quantified helps scientists, governments, and communities prepare for and respond to their devastating consequences—and adapt to a future where fire is becoming an ever more dominant force.

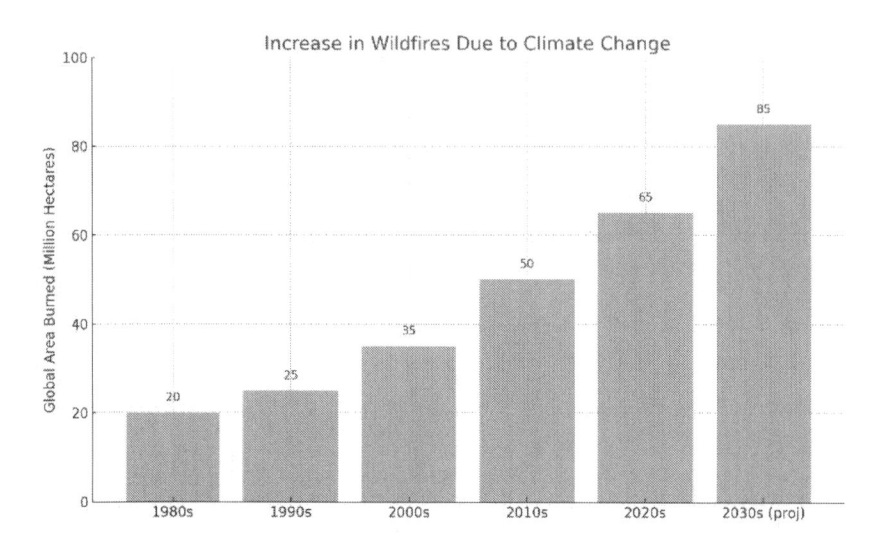

Increase in Wildfires Due to Climate Change

Decade	Global Area Burned (Million Hectares)
1980s	20
1990s	25
2000s	35
2010s	50
2020s	65
2030s (proj)	85

B. Examples of Wildfire Climate Adaptation

Climate adaptation to wildfires begins with understanding the evolving fire regime. Across regions such as the American West, the Mediterranean Basin, southeastern Australia, and boreal Canada, fire seasons are becoming longer, fires are burning hotter, and the wildland-urban interface (WUI) is expanding rapidly. Communities once considered safe are now regularly exposed to smoke, property destruction, and mass evacuations. Planners must, therefore, integrate climate projections with historical fire data to create bioregion-specific fire resilience strategies.

One prominent case is California's North Bay, where the 2017 Tubbs Fire and the 2018 Camp Fire devastated entire towns, such as Paradise. In response, the state established the Office of Wildfire Technology Research and Development and expanded funding for home hardening, defensible space, and prescribed burns. Yet even advanced states like California face equity issues: wealthier residents can rebuild with fire-resistant materials, while low-income renters and seniors often lack resources to return or adapt.

In Australia, the Black Summer fires of 2019–2020 led to a surge in national climate adaptation policy. Local governments began integrating fire-sensitive urban designs, such as green buffer zones and heat-resistant vegetation, into their planning codes. The town of Mallacoota, nearly cut off by fire, now invests in localized renewable energy microgrids and fire-resilient shelters. These innovations combine adaptation, decarbonization, and emergency preparedness.

In the Mediterranean, countries such as Greece, Spain, and Portugal have adopted satellite monitoring and AI-assisted fire risk prediction to pre-position emergency services and inform land-use zoning. Greece, following the deadly Mati wildfire in 2018, launched a national wildfire prevention strategy that incorporates public education, ecological firebreaks, and urban densification away from high-risk zones. However, implementation remains inconsistent, especially in informal settlements and agricultural zones where enforcement is weak.

Climate adaptation to wildfires must prioritize several key strategies. Land use policies should limit sprawl into high-risk zones and require the use of fire-resistant materials in all new construction. Zoning ordinances can encourage clustered development and create natural firebreaks through the use of greenbelts, orchards, or open spaces. Strategic relocation programs, though politically sensitive, may be necessary for communities to face repeated fire losses. Buffer zones can be enhanced through prescribed burns and Indigenous fire stewardship practices, which are increasingly recognized for their ecological efficacy and cultural importance.

C. Use of AI in Wildfire Adaptation and Response

Artificial intelligence is transforming how communities predict, prevent, respond to, and recover from wildfires. As wildfires intensify, increase in duration, and become more unpredictable due to climate change, AI provides powerful tools for early detection, risk assessment, emergency response, and long-term planning. The use of AI in wildfire management is not merely about automation; it is about amplifying human capacity to make life-saving, landscape-saving decisions faster and more effectively.

One of the most impactful uses of AI is real-time wildfire detection. Satellite-based systems, such as NASA's FIRMS (Fire Information for Resource Management System), now integrate AI algorithms that rapidly analyze thermal imagery, smoke plumes, and vegetation anomalies. Google's AI-powered wildfire boundary tracker uses satellite data to generate real-time maps of fire perimeters, accessible through Google Search and Maps, often hours before official updates.

In countries such as Chile, Spain, and the U.S., early detection, powered by machine learning, has reduced the time between ignition and response, enabling firefighters to mobilize sooner and protect more lives.

AI also supports predictive wildfire modeling, which is crucial for planning evacuations and resource deployment. Programs like IBM's Forecaster, Australia's Spark, and Canada's Prometheus simulate fire spread under varying wind, topography, humidity, and fuel load conditions. These platforms utilize historical fire data and meteorological inputs to forecast the likely path of fire and calculate damage probabilities across various time intervals. Such modeling helps emergency agencies pre-position firefighting aircraft, alert communities, and route evacuations more safely.

AI contributes significantly to vegetation risk mapping. Through remote sensing, LiDAR, and drone imagery, AI can classify vegetation types, assess moisture levels, and identify fuel loads. For instance, FireMap (used in California) integrates these data to prioritize fuel reduction treatments, such as prescribed burns, mechanical thinning, or grazing. AI not only detects where the danger lies —but it also suggests how to mitigate it efficiently.

In the wildland-urban interface (WUI), AI supports resilience planning. By analyzing urban growth trends, infrastructure layouts, and historic fire incidents, AI systems can highlight the most vulnerable neighborhoods and recommend zoning changes, fire-resistant materials, and defensible space standards. Municipalities can incorporate this data into comprehensive plans and code enforcement strategies. AI also helps insurers and real estate developers quantify fire risk, which could guide shifts in how and where buildings are constructed in the future.

AI is now playing a role in smoke forecasting and health adaptation. Programs like BlueSky in North America use AI to model smoke dispersion across multiple states or provinces. These forecasts help cities activate air quality response protocols, such as opening clean-air shelters, issuing advisories, or adjusting outdoor school activities to ensure the safety of students and staff. AI-enhanced air quality sensors, combined with crowd-sourced health data, are improving public health responses to long-distance smoke exposure.

AI also assists with community education and engagement. Platforms like Zencity and Pol. utilize natural language processing to analyze public concerns on social media and municipal surveys, enabling planners to communicate wildfire risks and adaptation strategies more effectively.

Emergency agencies are utilizing chatbots and virtual assistants to disseminate real-time evacuation instructions and respond to public queries during fire events.

Globally, AI is part of integrated climate adaptation strategies. In Portugal, for example, researchers have developed AI models that assess ignition probability and recommend land-use planning changes to reduce fire vulnerability. In India, AI is used to analyze changes in forest density and illegal land use that may increase fire risk. Meanwhile, in Indigenous fire stewardship programs, AI tools are being tailored to validate and amplify traditional knowledge about controlled burns and landscape health.

Despite these advances, ethical and governance challenges remain. AI models must be trained on high-quality, region-specific data to avoid biases that endanger underserved communities. Privacy concerns arise with the use of surveillance drones and real-time behavioral monitoring. Additionally, over-reliance on AI could sideline local expertise or community-based solutions.

D. Conclusion

AI is not a substitute for ecological wisdom, emergency services, or equitable planning. Instead, it is a powerful partner—an accelerant for resilience when grounded in justice, transparency, and science. As climate-fueled wildfires increase in scale and severity, AI can help humanity stay one step ahead, protect ecosystems, and build fire-adapted futures.

Advocacy Brief

Title: Burning Point: Wildfires, Climate Change, and the Urgent Need for Adaptation

Executive Summary

Climate change has transformed wildfires from seasonal events into escalating, year-round disasters. Fueled by hotter temperatures, prolonged droughts, earlier snowmelt, and increasingly volatile weather, wildfires now burn larger areas, spread faster, and devastate communities, ecosystems, and economies with unprecedented ferocity. This advocacy brief urges policymakers, planners, and community leaders to adopt bold adaptation strategies that mitigate wildfire risks through climate-responsive land management, early warning systems, infrastructure resilience, and environmental justice frameworks.

The Problem

Wildfires have become more destructive due to the compounding effects of global warming. According to the IPCC Sixth Assessment Report (2023), fire weather conditions—characterized by high temperatures, low humidity, and strong winds—have increased across many regions, including the western United States, Australia, southern Europe, and parts of South America and sub-Saharan Africa. The World Meteorological Organization (2024) reports a 30% increase in extreme wildfire events worldwide over the past two decades.

Longer fire seasons, drier vegetation, and stronger wind patterns create conditions that allow small ignitions to escalate into megafires quickly. The consequences include:

- Loss of human life and displacement of thousands
- Destruction of homes, infrastructure, and cultural landmarks
- Long-term ecological damage, including biodiversity collapse and soil degradation
- Severe air pollution from wildfire smoke causes spikes in respiratory and cardiovascular illnesses
- Increased carbon emissions that feedback into climate change worsen the crisis

Wildfires also disproportionately affect rural, Indigenous, and low-income communities, exposing them to environmental injustice through unequal preparedness, response, and recovery resources.

Policy Recommendations

1. Advance Climate-Responsive Land Management
Land-use policies must prioritize fire-adapted landscapes through controlled burns, mechanical thinning, and restoration of native fire-resistant vegetation. Indigenous fire stewardship practices—long suppressed—should be legally recognized and funded, as they offer proven models for maintaining ecosystem balance and preventing wildfires.

2. Mandate Wildfire Adaptation in Planning and Development
Municipalities and states should require Wildfire Impact Statements in all new developments in fire-prone areas, with mandatory buffer zones, non-flammable materials, defensible space regulations, and evacuation infrastructure. Zoning codes must integrate fire risk modeling into urban planning.

3. Expand Early Warning Systems and Real-Time Monitoring
Invest in AI-powered fire detection systems, satellite-based monitoring (e.g., NASA FIRMS), and sensor networks to provide early alerts and real-time updates. Increase public accessibility to Fire Weather Index ratings and localized risk dashboards to improve preparedness.

4. Protect Public Health

Governments must implement smoke-resilient infrastructure in schools, shelters, and public buildings, including air filtration systems and cooling centers, to ensure the safety of occupants. Health advisories and emergency services must be accessible to all communities, particularly those with limited transportation, language access, or digital connectivity.

5. Climate-Adaptive Infrastructure and Insurance Reform

Retrofit energy infrastructure to prevent fire ignitions (e.g., burying power lines, wildfire shutoff protocols). Reforming insurance markets to incentivize mitigation and to ensure post-fire recovery is equitable and climate resilient. Governments should establish public risk pools for uninsured or underinsured residents.

Call to Action

We urge immediate legislative action to fund wildfire adaptation, integrate Indigenous fire knowledge, and ensure all people—regardless of geography or income—are protected from the devastating effects of fire in a warming world.

Toolkit for Planners

This toolkit offers planning tools, policy models, data resources, and legal frameworks to assist communities in adapting to the intensifying wildfire threat exacerbated by climate change. It emphasizes prevention, preparedness, equity, and resilience.

1. Risk Assessment and Data Integration

Conduct Local Fire Risk Mapping

Use GIS-based wildfire hazard maps incorporating:

- *Vegetation types and fuel loads*
- *Slope and wind exposure*
- *Proximity to infrastructure and housing*
- *Fire history and future climate scenarios*

Tools:
- *CalFire FRAP Viewer*
- *USFS Wildfire Risk to Communities*
- *NASA FIRMS*

Integrate Climate Projections

Apply regionalized climate models to estimate future fire frequency, intensity, and season duration.

Source: IPCC (2023); NOAA Climate Toolkit; Local Climate Adaptation Frameworks

2. Land Use and Zoning Reforms
Establish Wildfire-Resilient Zoning

- *Restrict new development in Wildland-Urban Interface (WUI) zones*
- *Require minimum defensible space setbacks (30–100 ft)*
- *Prohibit flammable landscaping in fire-prone corridors*
- *Incentivize infill over expansion into fire-prone areas*

Require Wildfire Impact Assessments
Mandate Wildfire Impact Statements for all new subdivisions, utilities, and infrastructure, including:

- *Egress route analysis*
- *Water access for fire suppression*
- *Air quality and smoke shelter plans*
- *Environmental justice review*

3. Building and Infrastructure Standards
Retrofit and Code Upgrades

- *Require Class A fire-resistant roofing and non-combustible siding*
- *Install ember-resistant vents and window glazing*
- *Mandate rooftop sprinkler systems in high-risk zones*

Reference: California Building Code Chapter 7A; FEMA Wildfire Mitigation Guide

Utility Safety and Microgrid Planning

- *Bury power lines in extreme-risk corridors*
- *Support solar microgrids for backup power in evacuation centers and hospitals*
- *Require emergency water storage systems in new developments*

4. Community Engagement and Education
Support Local Fire Safe Councils
Fund and collaborate with neighborhood groups for:

- *Brush clearing and fuel breaks*
- *Prescribed fire coordination*
- *Evacuation drills and route planning*

Educating Home Hardening and Insurance
Launch public awareness campaigns in multiple languages with guidance on:

- *Retrofitting homes*
- *Applying for state/federal risk reduction grants*
- *Fire-safe landscaping and maintenance*

5. Health and Social Equity Integration
Design Smoke-Resilient Infrastructure

- Install HVAC air filters in schools, libraries, and shelters
- Establish clean air respite centers for vulnerable populations
- Include HEPA filtration in affordable housing upgrades
- Advance Equitable Recovery and Relocation
- Prioritize BIPOC and low-income communities in mitigation grants
- Create post-wildfire land trusts and managed retreat options with legal support
- Embed environmental justice analysis in fire adaptation planning

6. Funding and Policy Instruments
Secure Multi-Level Funding

- FEMA BRIC (Building Resilient Infrastructure and Communities)
- HUD CDBG-DR (Disaster Recovery)
- USDA Community Wildfire Defense Grants
- State Green Infrastructure Bonds
- Adopt Legal and Regulatory Mandates
- Incorporate wildfire mitigation in General Plans and Hazard Mitigation Plans
- Require Climate-Smart Development Checklists
- Enact local ordinances based on "Public Safety and Climate" nexus

7. Monitoring, AI, and Future Tools
AI-Based Fire Forecasting
Deploy predictive analytics to:

- Detect ignition points in real time
- Simulate fire spread under different wind and moisture conditions
- Guide evacuation orders and resource allocation
- Examples:
- IBM's FireCast
- Google AI wildfire maps
- International Bioregional AI Partnership (2025)
- Case Example: Ashland, Oregon's Community Wildfire Protection Plan (CWPP)
- Ashland integrated defensible space codes, urban-wildland buffer management, smoke planning, and support for low-income residents into its CWPP. It serves as a replicable model for mid-sized municipalities.

Resources

1. *Intergovernmental Panel on Climate Change (IPCC). (2023). AR6 Synthesis Report: Climate Change 2023.* https://www.ipcc.ch/report/sixth-assessment-synthesis-report
2. *World Meteorological Organization. (2024). State of Global Climate Report 2024.* https://public.wmo.int
3. *CalFire. (2024). Wildfire Resilience Strategy and Community Grants.* https://www.fire.ca.gov
4. *Cruz, M. J., & Kumar, S. (2024). Artificial intelligence in climate adaptation and wildfire response. Climate Intelligence Review, 11(2), 45–61.*
5. *Johnston, F. H., Bowman, D. M. J. S., & Henderson, S. B. (2024). Wildfire smoke and public health: Global evidence and responses. Environmental Health Perspectives, 132(1), 1–12.*
6. *Canadian Forest Service. (2023). Fire Weather Index System Overview.* https://cwfis.cfs.nrcan.gc.ca
7. *United Nations Office for Disaster Risk Reduction (UNDRR). (2024). Global Assessment Report on Disaster Risk Reduction.* https://www.undrr.org
8. *FEMA. (2023). Building Community Resilience to Wildfires: Wildland Urban Interface Toolkit. U.S. Federal Emergency Management Agency.* https://www.fema.gov
9. *USDA Forest Service. (2024). Wildfire Risk to Communities: Interactive Mapping and Planning Tool.* https://wildfirerisk.org
10. *Keeley, J. E., & Syphard, A. D. (2023). Fire intensity and forest mortality under climate extremes. Ecological Applications, 33(3), e2731.*
11. *United Nations Development Programme (UNDP). (2024). Climate Adaptation and Forest Fire Governance: A Rights-Based Approach.* https://www.undp.org

Chapter Seven
Earthquakes

Mami Mizutori, Special Representative of the UN Secretary-General for Disaster Risk Reduction (UNDRR):

> "Disasters are not natural. What turns hazards like earthquakes into disasters is vulnerability and lack of preparedness. Climate adaptation must include resilience to all risks—especially those that strike without warning."
> — *UNDER Global Platform, 2022*

Earthquake adaptation involves preparing societies, infrastructure, and ecosystems to minimize damage, save lives, and facilitate rapid recovery when seismic events occur. Unlike climate change-related adaptation, earthquake adaptation focuses on sudden-onset geophysical hazards. However, both require proactive planning, risk assessment, and community engagement. Climate change adaptation includes earthquakes because climate change impacts can worsen the effects of an earthquake.

A. What is an Earthquake?

An earthquake is a sudden and often violent shaking of the Earth's surface caused by the rapid release of energy stored in the Earth's crust. This energy travels in seismic waves that radiate from a focal point deep underground, often along geological faults. These faults are fractures in the Earth's crust where sections of rock have moved past each other. The most common cause of earthquakes is the movement of tectonic plates, the large sections of the Earth's lithosphere that constantly shift and interact with one another. When the stress from these movements exceeds the strength of the rocks involved, the resulting rupture leads to an earthquake.

Tectonic boundaries where earthquakes frequently occur include convergent boundaries, where plates collide; divergent boundaries, where they pull apart; and transform boundaries, where they slide past one another. Transform boundaries, like California's San Andreas Fault, are especially prone to significant seismic activity. In addition to tectonic causes, earthquakes can also be triggered by volcanic eruptions, landslides, and increasingly by human activities such as hydraulic fracturing (fracking), mining, and the impoundment of large reservoirs. These are known as induced earthquakes.

Earthquakes are widespread across the globe. Every year, the Earth experiences over 500,000 detectable earthquakes. Of these, about 100,000 are strong enough to be felt by people, and approximately 100 cause notable damage. Most earthquakes occur in well-known seismic zones such as the Pacific Ring of Fire, a vast region encircling the Pacific Ocean that includes parts of Asia, Oceania, and the western coasts of the Americas. Other seismically active areas include the Himalayan region, the East African Rift, and the Anatolian fault system in Türkiye.

The severity of an earthquake is measured in two key ways: magnitude and intensity. Magnitude refers to the amount of energy released at the source of the earthquake and is calculated using the Moment Magnitude Scale (Mw).

This scale has largely replaced the older Richter scale and is logarithmic, meaning that each whole-number increase represents approximately 32 times more energy release. For instance, a magnitude 7.0 earthquake releases 32 times more energy than a magnitude 6.0.

Intensity, in contrast, refers to the effects of an earthquake at specific locations, including damage to buildings and the sensation of shaking that people experience. This is often measured using the Modified Mercalli Intensity Scale, which ranges from I (not felt) to XII (destruction). Unlike magnitude, intensity varies by location depending on factors such as distance from the epicenter, depth of the earthquake, local soil conditions, and building design.

Major earthquakes can be catastrophic, especially when they strike densely populated or poorly prepared regions. Notable examples include the 2011 earthquake off the coast of Tohoku, Japan, which triggered a tsunami and nuclear disaster, and the 2023 earthquakes in Türkiye that caused widespread destruction and significant loss of life. In each case, it is not just the seismic energy that determines the disaster's impact but how healthy communities are prepared and how infrastructure is designed to withstand shaking.

Although scientists cannot predict earthquakes with absolute precision, seismic hazard models help identify regions at high risk. Some countries have deployed early warning systems that detect initial seismic waves and send alerts seconds before more damaging waves arrive—valuable time that can be used to shut down power systems, halt trains, or move people to safety. Technological advancements, including AI, satellite imaging, and dense sensor networks, are improving the speed and accuracy of these systems.

As urban populations grow and climate-related changes alter groundwater and land stability, more infrastructure is being built in seismically active areas. This makes proactive planning, seismic retrofitting, public education, and enforcement of building codes critical components of earthquake adaptation and resilience.

B. Global Urbanization and Earthquakes: A Growing Collision of Risk

As urbanization accelerates worldwide, more people are living in densely populated cities built in or near earthquake-prone regions. This convergence of rising urban density and seismic hazard represents one of the most pressing challenges for global disaster resilience. By 2050, it is projected that nearly 70% of the worldwide population will reside in urban areas, many of which are located along tectonic boundaries or in regions with inadequate enforcement of seismic building codes. The interaction between urban expansion and earthquake exposure heightens vulnerability—especially in the Global South, where infrastructure may be informal, aging, or unregulated.

Major cities such as Istanbul, Kathmandu, Mexico City, Jakarta, Tehran, and Port-au-Prince sit atop or near active fault lines. As these urban centers grow, they place increasing numbers of people, homes, schools, hospitals, and economic assets in harm's way. The problem is not just proximity to seismic zones but the quality of urban development. Informal settlements and slums—where millions live without adequate structural reinforcement—are particularly susceptible to collapse during moderate to strong quakes. In such environments, the same magnitude earthquake can result in exponentially greater casualties and damage compared to cities with earthquake-resistant infrastructure.

Urbanization also influences the secondary impacts of earthquakes. In dense cities, the effects of collapsed infrastructure can cascade: severed transportation routes delay emergency response, downed power lines cause fires, and ruptured water mains or sewage systems lead to public health crises. Moreover, the loss of housing and livelihoods following a significant earthquake in urban areas can lead to long-term displacement, economic instability, and social unrest.

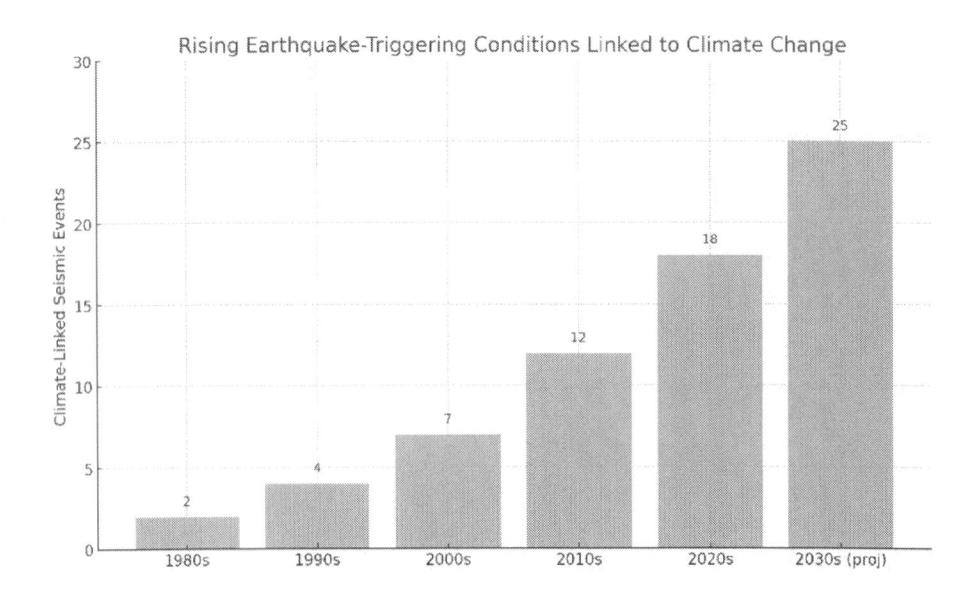

Rising Earthquake-Triggering Conditions Linked to Climate Change

Decade	Climate-Linked Seismic Events
1980s	2
1990s	4
2000s	7
2010s	12
2020s	18
2030s (proj)	25

Modern urbanization also changes the geophysical dynamics of seismic impact. Tall buildings, underground infrastructure, and land modifications can amplify ground shaking through a phenomenon known as site amplification, especially in sediment-filled basins. For example, Mexico City is built on a former lakebed that intensifies seismic waves, making even distant earthquakes highly destructive. Additionally, the urban heat island effect and groundwater extraction may weaken soils or increase the likelihood of liquefaction, which occurs when saturated ground loses its stability during shaking.

While some wealthy cities in seismically active areas—such as Tokyo, San Francisco, or Los Angeles—have implemented strict seismic codes, retrofitting programs, and early warning systems, many rapidly growing cities lack the necessary governance capacity, resources, or political will to do the same. The result is a growing global inequity in earthquake resilience, where the urban poor face the gravest risks but have the fewest protections.

To address this, integrated planning must prioritize earthquake-resilient urban design, enforcement of seismic codes, risk-informed zoning, and public education. Retrofitting vulnerable buildings, ensuring lifeline infrastructure (water, power, transit) is seismically robust, and decentralizing emergency response resources are key strategies. Moreover, mapping fault lines in urban peripheries and restricting high-density development in these zones can reduce future losses.

In summary, global urbanization is outpacing the world's seismic preparedness. As cities continue to expand into hazardous zones, the intersection of population density, fragile infrastructure, and seismic risk threatens to turn natural earthquakes into human disasters. Equitable, science-informed urban planning is essential to break this cycle and protect the billions now living on unstable ground.

C. Basic Climate Adaptation

Modern engineering solutions focus on seismic-resistant design. Buildings in earthquake-prone zones must meet stringent seismic codes. These structures utilize flexible materials, such as reinforced concrete and steel frameworks, and employ base isolation systems to absorb shock waves. Additionally, they retrofit older buildings with cross-bracing, shear walls, or dampers. Critical lifeline infrastructures, such as bridges, hospitals, and transportation networks, are also upgraded.

For instance, Japan's Shinkansen trains automatically stop when seismic activity is detected, and California's Bay Bridge was rebuilt with shock-absorbing piers after the 1989 Loma Prieta earthquake.
Urban and land use planning are essential components of earthquake adaptation. Governments implement zoning regulations and hazard mapping based on geological surveys, historical fault line data, and soil liquefaction risks. Cities like Tokyo and Santiago incorporate green and open spaces into their designs, serving both as daily public amenities and as emergency evacuation zones during disasters.

Community preparedness and education are equally vital. National early warning systems, such as J-ALERT in Japan and SASMEX in Mexico, provide public alerts seconds before strong shaking. Public schools and workplaces regularly conduct evacuation drills. Households are encouraged to maintain emergency kits with food, water, and first aid supplies and to follow home safety protocols, such as securing heavy furniture and learning how to shut off gas lines. Public education campaigns conducted through schools, community centers, and media outlets help ensure widespread awareness of earthquake safety protocols.

Governance and policy enforcement form the backbone of successful earthquake preparedness and response. Countries with strong regulatory frameworks, such as Japan and the United States, enforce strict building codes through construction inspections, penalties, and incentives for retrofitting. Insurance systems, such as the California Earthquake Authority and Japan's Japan Earthquake Reinsurance Company (JER), offer financial protection and encourage investments in adaptation. Post-disaster recovery strategies include legal mechanisms for emergency response, pre-positioned supplies, and plans for rapid reconstruction.

Japan is widely regarded as a global leader in earthquake preparedness and resilience. Following the 1995 Kobe earthquake, which caused over 6,000 deaths, Japan invested heavily in national seismic monitoring systems, smart city infrastructure, and local resilience centers. Chile, which experienced frequent seismic activity, responded to the 2010 Maule earthquake by strengthening public buildings, implementing mobile alert systems, and enhancing evacuation signage in coastal zones. Following the 2015 Gorkha earthquake, Nepal focused on retrofitting schools, preserving cultural heritage sites through UNESCO programs, and expanding mobile-based disaster preparedness education in rural regions.

Future directions in earthquake adaptation include the use of smart infrastructure embedded with AI sensors that detect tremors and automatically shut down critical systems. Cities are increasingly combining earthquake adaptation with climate and flood resilience strategies to form integrated urban resilience plans. Community co-design processes that involve marginalized groups are becoming central to effective planning, particularly in cities of the Global South. Countries such as Haiti, Afghanistan, and Indonesia continue to receive international support to enhance their earthquake resilience.

Recent references supporting these findings include the United Nations Office for Disaster Risk Reduction's Global Assessment Report on Disaster Risk Reduction 2025, the International Building Code for Seismic Design 2024 Edition published by the International Code Council, and the World Bank and GFDRR's 2024 report Seismic Resilience in Developing Cities: A Guide to Adaptation and Risk Reduction. The Japan Meteorological Agency's 2024 update on the Earthquake Early Warning System provides a detailed overview of national response technologies. Academic studies, such as Garschagen and Walsh's 2025 article on urban disaster governance, offer critical insights into best practices across global contexts.

D. Earthquakes and Climate Adaptation: Integrated Risk and Climate Adaptation Strategies for a Warming World

While earthquakes are not caused by climate change, they intersect powerfully with climate vulnerabilities in ways that demand integrated adaptation strategies. Many of the world's most densely populated urban centers are situated atop seismic fault lines and simultaneously face threats from climate-exacerbated risks, including flooding, extreme heat, water scarcity, and ecosystem degradation. In this era of cascading disasters, adaptation planning must be intersectional—addressing geophysical, climatic, ecological, and infrastructural vulnerabilities in tandem.

In regions with increasing exposure to both seismic activity and climate-related shocks, layered risk can lead to devastating outcomes. Climate change can weaken infrastructure, displace populations into geologically unstable zones, and increase the likelihood of landslides, dam failures, or soil instability—worsening the impacts of earthquakes. Meanwhile, communities recovering from earthquakes often face delayed rebuilding due to climate-related disruptions such as hurricanes, fires, or supply chain breakdowns.

Adaptation today must mean more than simply designing for one threat; it must encompass a broader range of considerations. It requires transforming planning, infrastructure, land use, and governance to create systems that are resilient across a spectrum of hazards.

E. Integrated Approaches: Land Use, Construction, and Nature-Based Adaptation

Integrated planning begins with identifying geographic zones that overlap in hazards. Cities must move away from siloed zoning and instead adopt multi-hazard maps that simultaneously assess earthquake fault lines, floodplains, landslide-prone hillsides, subsidence zones, and urban heat islands. These maps must be continuously updated using high-resolution data and shared across planning, transportation, and emergency departments.

Infrastructure retrofitting should also incorporate both earthquake resistance and climate resilience. For instance, a school retrofitted with base isolators for seismic protection can also be elevated to avoid flood damage and equipped with solar panels to maintain operations during grid failures. Water systems can be designed to both resist quake ruptures and operate during droughts. Cooling green roofs can also serve as safe open spaces in the aftermath of seismic shocks.

Nature-based solutions offer dual protective benefits. Mangroves, wetlands, and reforested slopes mitigate the impacts of storm surges and flooding while also stabilizing soil and absorbing seismic energy. In earthquake zones where landslides are a risk, reforesting degraded hills can mitigate both climate and tectonic erosion.

F. Expanded Use of Artificial Intelligence for Earthquake and Climate Adaptation

Artificial Intelligence (AI) revolutionizes our understanding, prediction, and response to the intersection of seismic and climate risks. AI models can now integrate data from satellite imagery, seismic sensors, hydrological stations, population movements, and building footprints to create dynamic risk profiles that evolve in real time.

In earthquake zones, AI systems trained on historical seismic data and fault stress maps can now forecast the probability of aftershocks and simulate building collapses under different quake scenarios.

When layered with climate data—such as precipitation trends, drought conditions, and aquifer depletion, these models identify areas where soil instability or liquefaction is amplified due to climate stress. AI also aids groundwater stress analysis, a critical issue in many seismic-prone regions where over-extraction causes land subsidence and structural vulnerability.

Visualizations generated by AI can provide powerful decision-support tools. Urban planners and emergency responders can utilize AI-generated 3D simulations to model cascading events, such as a scenario in which an earthquake ruptures gas lines during a heatwave or a landslide blocks flood evacuation routes after a tremor. These compound risk scenarios are vital for preparedness and insurance planning.

AI-powered drones and computer vision are also used to rapidly assess post-earthquake damage in areas cut off from immediate access. AI can identify cracked buildings, broken water mains, and collapsed roads within hours, even during storms or poor visibility, thereby accelerating both rescue and reconstruction efforts. Importantly, such technologies are increasingly being used in the Global South, where traditional monitoring infrastructure may be lacking, but mobile-based AI offers a leapfrogging opportunity.

G. Expanded Case Studies of Earthquake and Climate Adaptation

Japan provides one of the most advanced models of integrated disaster resilience. As one of the most earthquake-prone countries on Earth and also vulnerable to typhoons, sea-level rise, and aging infrastructure, Japan has invested heavily in multi-hazard adaptation. Tokyo's Metropolitan Resilience Plan features floodgates that also serve as earthquake evacuation platforms. These vertical parks serve as both green cooling infrastructure and seismic assembly zones, as well as underground reservoirs reinforced to withstand both hydraulic pressure and seismic waves. Japan's disaster parks, such as the Tokyo Rinkai Disaster Prevention Park, serve as education hubs and emergency coordination centers designed for both earthquake and extreme weather events.

Mexico City faces severe challenges related to both climate and earthquakes. Built on a dried lakebed, the city experiences significant subsidence, which is compounded by groundwater over-extraction intensified by climate-induced drought. When earthquakes strike, buildings in this softened soil experience amplified shaking.

Recent adaptation efforts have focused on seismic retrofitting of public schools, the creation of urban rain gardens to reduce subsidence, and the use of AI modeling to track water levels, soil movement, and structural vulnerability. Pilot projects led by universities are utilizing AI to correlate earthquake damage zones with aquifer depletion patterns, yielding new insights into infrastructure risk in a warming, drying world.

California exemplifies how climate and earthquake risks can intersect with wildfire, housing crises, and water stress. In Los Angeles and San Francisco, adaptation plans now explicitly integrate earthquake early warning systems with heat alert protocols and flood mitigation strategies. Several counties have launched "Resilience Hubs" that combine seismic retrofitting, solar microgrids, and cooling centers in vulnerable neighborhoods. AI is being used to map retrofitting priorities based on building age, seismic hazard, projected wildfire spread, and social vulnerability indices, ensuring that equity is embedded in resilience investments.

Turkey offers a compelling example following the 2023 Kahraman Maras earthquakes. The rebuilding process now incorporates climate-aware design, such as elevated shelters to avoid future flood risks and shaded refugee housing to mitigate extreme summer heat.

AI is being deployed to match housing needs with safe, climate-resilient land parcels, taking into account soil stability, projected heat increases, and access to water resources.

Indonesia, facing both seismic risk from the Ring of Fire and climate threats such as sea-level rise and tropical storms, has begun relocating its capital to a safer site in Borneo. The new capital, Nusantara, is being designed with earthquake-resistant buildings, green infrastructure for flood absorption, and AI-optimized layouts that place critical facilities away from combined hazard zones. This is among the world's first large-scale examples of proactive relocation based on both seismic and climate forecasts.

H. Conclusion: Toward Multi-Hazard, Equitable, AI-Informed Climate Adaptation

The intersection of earthquakes and climate change demands a profound shift in how we define adaptation. No longer can communities afford to plan for one disaster at a time. The future will be determined by compounding risks and the tools we use to adapt must be equally compound. Artificial Intelligence, when used ethically and equitably, offers unprecedented potential to model, visualize, and act on these complex systems of vulnerability.

Adaptation planning must include legal mandates, land use controls, cross-sector data sharing, Indigenous knowledge, public education, and digital tools that enable decentralized and community-informed decision-making. In this context, AI is not simply a high-tech solution—it is a framework for seeing connections, forecasting harm, and designing a safer, greener, and more resilient world.

Advocacy Brief

Title: Strengthening Cities, Saving Lives: Climate Adaptation for Earthquake Resilience
The Challenge
As climate change accelerates urbanization and environmental stress, earthquake risks are intensifying—especially in megacities located along active fault lines. Growing populations, informal housing, aging infrastructure, and climate-compounded land instability (e.g., drought, groundwater depletion, glacial melt) all amplify the vulnerability of urban regions to seismic disasters.
Why Action Is Urgent

Each year, over 100 significant earthquakes occur globally, many in areas unprepared for the cascading impacts: collapsed buildings, disrupted lifelines, mass displacement, and long-term economic trauma. In a warming world, the combination of seismic hazard and urban exposure is a recipe for catastrophic loss —unless we act now.
Key Adaptation Priorities

1. *Strengthen Building Codes and Retrofitting: Mandate and enforce seismic safety standards for all new construction and prioritize retrofitting of schools, hospitals, and housing in vulnerable zones.*
2. *Resilient Urban Planning: Integrate seismic hazard mapping into zoning and development decisions to enhance resilience. Avoid dense construction in high-risk zones and enforce land-use regulations.*
3. *Early Warning and Public Education: Invest in AI-enhanced early warning systems, evacuation planning, and earthquake drills in schools and workplaces to enhance safety and preparedness.*
4. *Equity in Preparedness: Direct funding and technical support to low-income, marginalized, and informal communities at highest risk.*

Call to Action

We urge governments, city planners, and international donors to make earthquake adaptation a core component of climate resilience strategies. Lives, livelihoods, and entire cities depend on proactive, inclusive seismic planning.

Planners Toolkit

Integrating Seismic Resilience and Climate Change Preparedness in Urban Planning

I. Introduction: Why Integration Matters

Municipalities worldwide are increasingly facing compound disasters—such as heatwaves followed by earthquakes, drought-induced subsidence weakening building foundations, or floods impairing emergency response to seismic events. While earthquakes are not directly caused by climate change, their impacts are exacerbated by climate-induced vulnerabilities, including deteriorating infrastructure, stressed ecosystems, population displacement, and land instability. This toolkit provides local governments with a roadmap for integrating earthquake resilience into climate adaptation strategies anchored in science, equity, AI technologies, and community-based planning.

II. Foundational Principles for Municipal Integration

- *Multi-Hazard Planning: Replace siloed emergency plans with unified strategies that assess seismic, hydrological, meteorological, and environmental risks together.*
- *Climate-Adjusted Seismic Zones: Update building codes and land use zones based on new data showing how drought, groundwater depletion, and temperature rise affect soil behavior during earthquakes.*
- *AI and Predictive Modeling: Use artificial Intelligence to assess risk in real-time and inform resource allocation, infrastructure retrofits, and evacuation planning.*
- *Justice-Centered Resilience: Prioritize retrofitting, early warning systems, and green infrastructure in historically marginalized and hazard-exposed communities.*

III. Planning Actions for Municipal Governments

1. Urban Infrastructure Resilience

- *Retrofit public buildings (schools, hospitals, emergency shelters) for both seismic and climate risks.*
- *Require earthquake- and flood-resistant designs in new construction through updated municipal codes.*

- *Install decentralized utilities, such as solar-powered microgrids, rainwater catchment systems, and flexible water pipes.*
- *Elevate critical roads and bridges in seismic zones that are also susceptible to flooding or subsidence.*

2. Land Use and Zoning

- *Implement multi-hazard risk zoning maps integrating climate projections (IPCC AR6) and seismic fault models.*
- *Prohibit high-density development on liquefaction-prone soils and unstable slopes worsened by drought.*
- *Establish green buffer zones—such as wetlands, bioswales, and tree corridors—that absorb seismic energy and mitigate climate shocks.*

3. AI and Digital Tools

- *Use AI-enhanced 3D hazard simulations for public education, permitting, and emergency planning.*
- *Incorporate real-time data from sensors, drones, and satellites to detect stress on critical infrastructure.*
- *Develop digital twin models of neighborhoods to simulate building collapse, evacuation, and climate hazards in combined scenarios.*

4. Community Resilience and Public Health

- *Design and support Resilience Hubs in neighborhoods as multi-use facilities equipped with cooling stations, seismic shelters, microgrids, and first aid supplies.*
- *Expand early warning systems that integrate both earthquake alerts and extreme weather notifications.*
- *Train residents through neighborhood-based disaster simulations and hazard mapping exercises to enhance their preparedness.*
- *Invest in trauma-informed disaster recovery plans to address mental health impacts after earthquakes or climate disasters.*

IV. Financing and Legal Mechanisms

- *Establish a Municipal Climate-Resilience Seismic Fund utilizing bonds, insurance savings, federal grants, and carbon revenue.*
- *Mandate Personal and Commercial Environmental Impact Statements for significant developments, including seismic and climate exposure.*

- Use zoning incentives, green building tax credits, and resilience loans to fund retrofits and climate-smart construction.
- Require mandatory disclosure of seismic and flood risks in all real estate transactions and rental agreements.
- Reduction in recovery time after compound events (baseline vs. post-adaptation)
- Equity Index: Degree to which adaptation investment reaches high-vulnerability zones

Next Steps:

1. Conduct a Municipal Hazard Convergence Audit using AI mapping.
2. Form a Resilience Integration Task Force comprising representatives from planning, public health, utilities, and community leaders.
3. Apply for multi-hazard adaptation grants via national climate resilience or hazard mitigation programs.
4. Engage the public through interactive visualization tools and multilingual community drills.
5. Publish a Resilience Transparency Report annually with metrics, updates, and adaptation outcomes.

Resources

1. *(2023). Sixth Assessment Report: Impacts, Adaptation, and Vulnerability. Intergovernmental Panel on Climate Change.*
2. *(2024). Earthquake Hazards and Climate Linkages. United States Geological Survey.*
3. *World Bank. (2025). Climate Resilient Cities: Integrating Earthquake and Climate Risk.*
4. *(2024). Global Disaster Risk Reduction Framework for Local Governments.*
5. *California Office of Emergency Services (Cal OES). (2024). Integrated Climate-Earthquake Resilience Planning Guide.*
6. *United Nations Office for Disaster Risk Reduction (UNDRR). (2024). Global Assessment Report on Disaster Risk Reduction. https://www.undrr.org*
7. *U.S. Geological Survey (USGS). (2024). Earthquake Hazard Maps and Urban Seismic Risk Framework. https://www.usgs.gov*
8. *Federal Emergency Management Agency (FEMA). (2023). Planning for Seismic Resilience: Community Adaptation Toolkit. https://www.fema.gov*
9. *World Bank. (2024). Building Resilient Cities: Earthquake Risk and Climate-Integrated Infrastructure Planning. https://www.worldbank.org*

10. Bilham, R., & Wallace, K. (2024). *Urban Earthquake Risk in the Climate Century.* Seismological Research Letters, 95(2), 145–163.
11. Cruz, M. J., & Kumar, S. (2024). *AI-Enhanced Early Warning Systems for Earthquakes and Cascading Hazards.* Climate Intelligence Review, 11(1), 23–40.
12. National Institute of Building Sciences (NIBS). (2024). *Mitigation Saves: Seismic Adaptation and Co-Benefits Under Climate Stress.* https://www.nibs.org
13. Global Earthquake Model Foundation (GEM). (2023). *Global Seismic Risk Map and Adaptation Guidance.* https://www.globalquakemodel.org
14. United Nations Development Programme (UNDP). (2024). *Inclusive Earthquake Adaptation and Risk Governance.* https://www.undp.org

Chapter Eight

Designing for Climate Adaptation: The Role of Place

"The law tends to treat environmental harms in isolation, but climate change makes that impossible. We need cumulative impact frameworks that reflect how people actually live—in real, complex places."
— *Catherine L. O'Neill, Seattle University School of Law, speaking at the Environmental Justice and Law Symposium (2024)*

A. Why Place Studies Are Better for Climate Adaptation and Resilience

Localized Knowledge and Relevance
Place Studies centers the unique geographic, ecological, and cultural characteristics of specific locations. In contrast to generic or one-size-fits-all approaches, climate adaptation guided by Place Studies draws on local knowledge, lived experience, and site-specific conditions. This makes strategies more accurate, effective, and trusted. For example, coastal communities, arid zones, mountainous regions, and urban heat islands each face different vulnerabilities. Place-based adaptation ensures solutions are tailored to real local risks—floodplain planning, water harvesting, or heat shelter design—rather than imposed from a distance.

Integration of Cultural and Ecological Histories

Climate resilience isn't just about physical infrastructure, it's about people, identity, and place. Place Studies allows adaptation efforts to reflect the deep histories of how communities have interacted with the land, water, and climate over time. This is especially vital in Indigenous and frontline communities where cultural survival is tied to specific places. By restoring relationships with land, honoring ancestral stewardship, and reactivating traditional ecological knowledge (TEK), Place Studies build not just resilience but cultural renewal.

Holistic Understanding of Systems and Stressors

Place Studies fosters an integrated view of social, ecological, and economic systems. Instead of isolating climate from housing, health, water access, or transportation, it sees them as interconnected. This systemic view is essential for climate adaptation because it recognizes cumulative impacts and feedback loops —like how drought can affect migration, food security, and mental health at once. Place Studies helps practitioners and policymakers plan across sectors to prevent cascading failures during climate disruptions.

Community Agency and Equity

Adaptation that emerges from Place Studies inherently values community voices. It shifts power away from top-down technocratic models and toward participatory, democratic planning processes.

This builds social resilience by including those who are most affected—often low-income, BIPOC, or Indigenous communities—in decision-making. In turn, this increases the legitimacy, relevance, and uptake of adaptation strategies, while addressing historical injustices that made certain communities more vulnerable in the first place.

Flexibility and Iterative Design

Because Place Studies is rooted in the specificity of each locale, it encourages adaptive management—policies and practices that can evolve in response to real-time conditions. This is key in a rapidly changing climate where predictions may shift. Place-based resilience practices such as watershed-based planning, living shorelines, or culturally tailored early warning systems can be tested, refined, and scaled as appropriate.

Identification of Root Causes and Structural Risk

Place Studies reveals the layered history of land use, zoning, industrial development, and racialized policies that created existing environmental risks. This historical lens is crucial for climate resilience because it helps identify why some areas flood repeatedly, why others suffer heat burdens, and why certain populations are stuck in vulnerable zones. Addressing these systemic roots is more transformative than merely reacting to surface-level climate effects.

Supports Bioregional Governance

Place-based adaptation aligns with the concept of bioregionalism—organizing governance and planning around ecological regions rather than political boundaries. Rivers, forests, watersheds, and migration corridors don't respect county lines. Place Studies encourage cross-jurisdictional planning that is responsive to ecological realities, helping regions adapt collectively to climate pressures like wildfires, sea-level rise, or biodiversity loss.

Strengthens Social Cohesion and Mental Health

Resilience isn't only technical—it's emotional and social. Place Studies reinforces identity, belonging, and meaning by connecting people to their environment. This helps buffer against climate anxiety, displacement trauma, and fragmentation. When people are engaged in place-making, restoration, and storytelling, they build a sense of collective purpose that strengthens psychological resilience.

Fosters Innovation through Local Experimentation

Place Studies supports the development of local climate adaptation labs, living laboratories, and community pilot programs. Because strategies are tailored to the physical and cultural environment, they are more likely to succeed and be replicated elsewhere with contextual adjustments. These hyperlocal efforts become innovation engines for regionally appropriate technologies, land management, and social models.

Bridges Science, Policy, and Community Wisdom

Lastly, Place Studies bridges the divide between data-driven science and community-based wisdom. It encourages transdisciplinary collaboration between climatologists, urban planners, educators, Indigenous leaders, artists, and youth. This convergence of knowledge systems produces deeper, more inclusive, and enduring climate adaptation solutions—rooted in the soil, story, and spirit of place.

B. Designing for Climate Adaptation in Place

Adaptation labs are emerging around the world as creative, place-based learning spaces that empower youth to design for a changing climate. These labs take many forms—mobile classrooms, innovation hubs, field-based programs, or school-integrated maker spaces—but they all share a common mission: to equip communities, especially young people, with the tools, knowledge, and imagination needed to adapt to the environmental challenges of today and tomorrow.

C. Examples of Place Based Climate Adaptation

In **New York City**, the Harlem Resilience Hub has engaged high school students in mapping areas at risk of urban flooding and heat exposure using drones and GIS software. Students proposed converting unused alleyways into shaded green spaces with water-absorbing bioswales and benches made from recycled materials. Their work informed city planning proposals and showcased how local youth insights can influence real infrastructure.

In **New Orleans**, students in the Waterwise Neighborhood Lab design solutions to chronic flooding in their own neighborhoods. They've created models for raised garden beds that double as stormwater management tools and have partnered with engineers to pilot permeable pavement projects. This lab is rooted in the unique ecological and social challenges of living in a low-lying delta city prone to hurricanes and sea level rise.

In **Navajo Nation**, Indigenous youth participate in adaptation labs that combine traditional ecological knowledge with climate science. Projects focus on restoring dryland farming techniques, reviving ancestral water harvesting methods, and using storytelling to map environmental changes observed over generations. These labs are not only technical; they are cultural spaces where language, memory, and stewardship converge in climate education.

In **Bangladesh**, where climate displacement due to sea level rise and flooding is a daily reality, Floating School Labs use boats outfitted with solar panels, Wi-Fi, and lab equipment to bring climate education to remote rural students. Young people conduct experiments on water salinity, test filtration methods, and brainstorm climate-resilient housing prototypes for flood-prone regions. These floating labs are critical for both access to education and adaptation planning.

In **Los Angeles**, the Youth Climate Action Lab has partnered with architecture students and city planners to design cooling infrastructure in asphalt-heavy schoolyards. Projects include tree planting campaigns, installation of misting stations, and shade structure design using low-cost bamboo and recycled textiles. These designs address both the urban heat island effect and environmental equity, as many underserved schools lack green space.

In **South Africa**, the Green Futures Adaptation Academy trains young people in agroecology, eco-construction, and water conservation. Participants learn to build cob structures that withstand climate extremes, restore native vegetation, and implement greywater reuse systems. The academy prepares youth for green jobs while directly improving local resilience through practical, land-based interventions.

These examples show the wide variety and power of adaptation labs. Whether situated in coastal villages, urban neighborhoods, deserts, or floodplains, they all connect youth to place, empower them to solve problems with creativity and care, and weave resilience into the land use, food systems, and infrastructure of their communities. By expanding these labs and supporting their long-term integration into education and planning systems, we invest in a future shaped by informed, imaginative, and deeply grounded climate leaders.

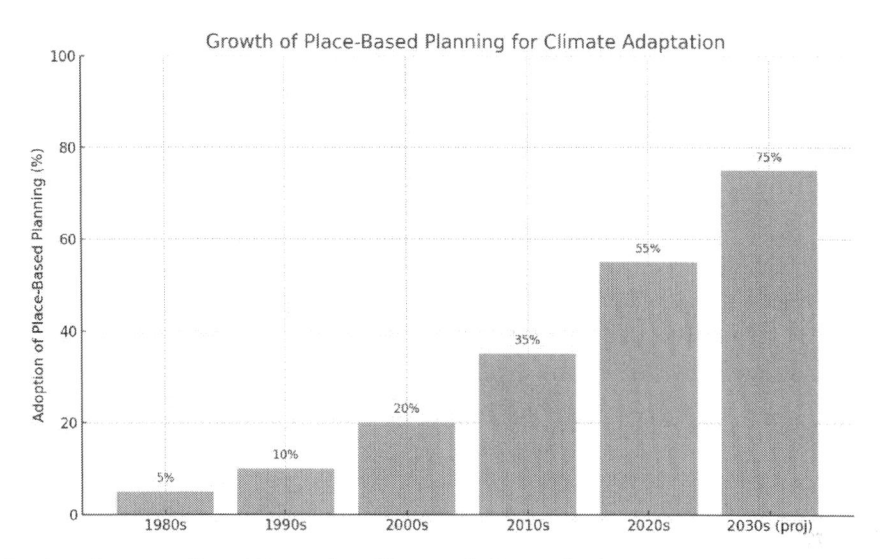

Growth of Place-Based Planning for Climate Adaptation

Decade	Adoption of Place-Based Planning (%)
1980s	5%
1990s	10%
2000s	20%
2010s	35%
2020s	55%
2030s (proj)	75%

D. Understanding Place Studies and Cumulative Impacts

Place Studies is an interdisciplinary field that explores the relationships between people, culture, environment, and specific geographic locations. It emphasizes the ways in which landscapes are not just physical spaces but are shaped by histories, social dynamics, politics, and ecological processes. In the context of environmental justice, Place Studies provides a framework for understanding how marginalized communities often bear disproportionate environmental burdens and how their identities and experiences are rooted in specific, often exploited, places.

Place-Based Justice and Environmental Law

In Oregon, Place Studies became particularly significant in legislative and legal contexts through the work of scholars and advocates like Robert W. Collin. I applied Place Studies to highlight the intersection of race, geography, and environmental law. He argued that environmental harm is not randomly distributed, often concentrating in communities of color and low-income neighborhoods. This advocacy helped shift policy discussions in Oregon toward recognizing "place" as central to understanding and addressing cumulative environmental impacts.

Place Studies and Indigenous Land Stewardship

Drawing on Place Studies, I also uplifted Indigenous knowledge and land stewardship practices in his Oregon testimony. He cited tribal ecological knowledge and the importance of tribal sovereignty in land use planning, particularly in rural Oregon and areas affected by forestry, water rights disputes, and extractive industries. I urged legislators to treat Indigenous communities not as stakeholders but as sovereign governments with deeply rooted relationships to place.

Legal Frameworks and Land Use Reform

Through Place Studies, Collin advocated for structural reforms to Oregon's land use system. I criticized the ways in which urban growth boundaries, industrial zoning, and lack of community engagement perpetuated environmental injustice. His testimony pushed for inclusive land use reform that integrated environmental justice screening tools, ecological zoning, and democratic planning processes that reflect the lived experiences of those most impacted. Ecological zoning would include cumulative environmental impacts of pollution.

E. Examples of Place Based Climate Adaption and Place Based Planning

In Harlem, New York City, community-led initiatives and researchers have used place-based tools to map the cumulative impacts of urban heat, air pollution, and disinvestment. Through GIS mapping and participatory research, youth and local organizations documented how formerly redlined neighborhoods—predominantly Black and Latino—experience higher temperatures, fewer green spaces, and increased rates of asthma. These overlapping stressors are not accidental; they are products of layered policies of racial exclusion, zoning, and underfunding. Using this data, advocates successfully pressured city planners to install cooling infrastructure, increase tree canopy coverage, and revise emergency response protocols, proving that place-based and cumulative impact approaches can directly inform climate justice policy.

The Biloxi-Chitimacha-Choctaw tribe in coastal Louisiana illustrates a powerful case of place-based cumulative vulnerability. Centuries of land loss from levee construction, oil canal dredging, saltwater intrusion, and rising sea levels culminated in their forced climate relocation. Place Studies allowed tribal members and researchers to trace the cumulative degradation of wetlands and the intergenerational loss of food systems, sacred sites, and community cohesion. Rather than relocating with a generic housing plan, the tribe insisted on culturally grounded design rooted in ancestral land relationships. Their story underscores how cumulative impacts—ecological, historical, and social—demand adaptation strategies that respect place, memory, and sovereignty.

Legal scholar Robert W. Collin's testimony in Oregon stands as a foundational example of using Place Studies to reveal cumulative environmental harm. In North Portland's historically Black Albina neighborhood, Collin and collaborators showed how industrial zoning, freeway construction, and neglect of public services created a landscape of cumulative risk—exposing residents to diesel particulate matter, soil contamination, and noise pollution. His place-based analysis and public testimony influenced Oregon's creation of the Environmental Justice Task Force and shaped state efforts to incorporate environmental equity into land use planning, showing how cumulative impact assessments grounded in place can shape legislative reform.

In the fire-prone regions of British Columbia, Indigenous communities such as the Tsilhqot'in have restored traditional fire practices as a place-based response to cumulative impacts from colonial fire suppression, industrial logging, and climate-driven drought. Through Place Studies, they've documented how the loss of Indigenous fire knowledge disrupted ecological balance, increased fuel loads, and reduced biodiversity. By reintroducing cultural burning, these communities are not only restoring ecological health but also addressing the cumulative effects of land mismanagement, policy erasure, and cultural marginalization. Their success is now informing provincial fire strategies.

In the Mathare informal settlement in Nairobi, Kenya, cumulative risks include flood exposure, poor sanitation, unregulated housing, and contaminated drinking water. Youth-led organizations used place-based assessments to document how these risks intersect—where sewage backs up into homes during storms, mosquito-borne illnesses surge, and emergency services fail to reach narrow, unplanned alleys. Their participatory mapping and storytelling efforts transformed these overlooked areas into visible climate risk zones. As a result, climate adaptation strategies—such as permeable pavement and decentralized water storage—are now being piloted, illustrating how place-specific cumulative data can reshape infrastructure priorities.

In coal-impacted areas of Central Appalachia, Place Studies have helped local residents document the cumulative impacts of mountaintop removal mining—loss of biodiversity, water contamination, deforestation, and increased cancer rates. Residents mapped dead streams, conducted well water testing, and collected oral histories to make the landscape of sacrifice visible. These place-based studies formed the foundation for lawsuits, state regulatory challenges, and health studies. By rooting resilience efforts in community knowledge, Appalachia's adaptation strategies now include ecological restoration, renewable energy projects, and land buyback cooperatives—demonstrating resilience as both ecological repair and social justice.

Mexico City faces cumulative challenges from aquifer over-extraction, loss of wetlands, unchecked urban sprawl, and historical erasure of Indigenous water management. Place Studies in Xochimilco and the former Lake Texcoco basin have documented how urban design disconnected the city from its aquatic identity. This has led to land subsidence, urban flooding, and water inequity. By reviving chinampas (floating gardens), re-greening canals, and integrating community-led hydrological education, adaptation efforts are now addressing not just the symptoms (flooding and water scarcity) but the cumulative, place-specific degradation of a once-thriving ecosystem.

In Yellowknife and surrounding Indigenous territories, Place Studies are used to track the cumulative impacts of permafrost thaw, climate-induced displacement of caribou, and colonial policies that disrupted traditional food networks. Dene and Métis youth are leading permafrost monitoring projects using a combination of drone technology, oral history, and traditional navigation routes. These place-based programs help communities adapt while addressing cumulative losses in land access, cultural transmission, and ecological function. The data is also shaping federal climate investments in northern infrastructure and food sovereignty.

In the Sundarbans and southern delta of Bangladesh, Place Studies have revealed how deforestation, embankment failures, cyclones, and salinization have compounded vulnerability in fishing and farming communities. Community groups have responded with floating gardens, saline-resistant crops, and schoolboats—all rooted in local design and traditional adaptation practices. These efforts address not just rising waters but the cumulative erosion of livelihoods, migration pressures, and governance failures. Their success has influenced the UN and global climate adaptation funders to prioritize place-based, community-led adaptation.

In the agricultural heart of California, migrant farmworkers—predominantly from Latinx and Indigenous communities—face cumulative risks from extreme heat, pesticide exposure, poor housing, and lack of healthcare. Place Studies conducted by UC researchers and local groups have combined temperature tracking, toxic drift mapping, and oral testimonies to reveal how labor exploitation intersects with ecological degradation. These findings have prompted policy changes including mandatory heat protection standards, pesticide buffer zones near schools, and emergency cooling centers in rural towns. Place-based analysis here directly informed both labor and environmental reform

E. AI and Placed Based Planning

Artificial Intelligence (AI) and Place-Based Climate Adaptation may seem like approaches from opposite ends of the spectrum—one data-driven and global, the other rooted in local knowledge and specific geographies. Yet their integration offers a powerful convergence: AI can process complex climate data at unprecedented scales, while place-based adaptation ensures that interventions are meaningful, just, and grounded in community realities. When paired thoughtfully, AI can amplify the effectiveness of place-based strategies by translating hyperlocal observations into predictive models, resource maps, and decision-making tools tailored to the unique contours of place.

Cumulative environmental impacts—caused by overlapping stressors such as pollution, heat, flood risk, and social vulnerability—are inherently place-based. AI can synthesize large, multi-source datasets (e.g., satellite imagery, sensor data, census demographics, health outcomes) to map cumulative risk zones with high spatial and temporal resolution. For example, machine learning algorithms have been used in Portland, Oregon, to model air quality disparities and overlay them with historical redlining maps, identifying neighborhoods facing environmental injustice. These AI-generated insights support targeted adaptation interventions that align with the lived experience of residents.
AI-driven climate models are increasingly being customized to specific geographies. Deep learning algorithms can downscale global climate data to reflect local terrain, land use, and microclimates—key variables in place-based adaptation. In coastal communities, AI-powered early warning systems use tide gauges, satellite altimetry, and local rainfall data to predict flooding street-by-street. In arid bioregions, AI helps map groundwater recharge zones and project drought impacts with regional specificity, improving agricultural resilience and water conservation planning.

Place-based adaptation depends on community engagement and ground-truthing. AI systems can incorporate participatory data from citizens using smartphones, sensors, and mapping apps to track environmental changes—like temperature spikes, flood events, or air quality alerts. In Nairobi's Mathare Valley and parts of India's Ganges Basin, AI tools combined with community-collected data are used to identify informal settlement vulnerabilities. These AI-enhanced participatory platforms democratize adaptation by recognizing residents as producers—not just consumers—of climate knowledge.
AI can support—not replace—Indigenous and ancestral place-based knowledge. When developed with cultural sensitivity and data sovereignty, AI models can help validate and scale traditional ecological practices such as controlled burns,

seasonal water management, or biodiversity monitoring. Projects in northern Australia, Canada, and the Amazon have begun integrating Indigenous place names, ecological indicators, and oral histories into AI-powered conservation and adaptation platforms. Ethical AI design ensures these systems respect intellectual property rights and reinforce community agency.

Planners and policymakers increasingly use AI-based tools to simulate future land use scenarios based on place-specific climate risks. For example, AI models can predict where sea-level rise will intersect with affordable housing stock or where wildfire risk overlaps with transit-dependent populations. In California and the Netherlands, AI platforms inform zoning adjustments and infrastructure investments that support long-term community resilience while avoiding climate gentrification.

AI has proven vital in place-based climate-health adaptation. In the U.S. Midwest and parts of Southeast Asia, AI systems use weather patterns, mosquito breeding data, and population movement to forecast dengue and malaria outbreaks. These systems enable timely interventions in vulnerable areas—spraying, education, or medical resource allocation—down to the neighborhood scale. AI also models urban heat exposure and identifies elderly populations or outdoor workers at greatest risk, aligning public health adaptation with precise place-based strategies.

During climate-induced disasters—floods, wildfires, heatwaves—AI processes real-time data to improve local emergency response. NLP tools analyze social media and emergency call transcripts to identify distress locations. AI-powered drones and satellite imagery assess damage and accessibility, guiding first responders in geographically complex settings. Importantly, localized training data improves AI accuracy: a heat vulnerability map for Phoenix, Arizona, must differ from one for Kolkata, India.

AI can track adaptation not just through environmental indicators but also through human well-being—aligned with flourishing frameworks. Using place-based inputs (e.g., access to green space, food security, education quality, safety), AI can help governments measure resilience in terms of thriving, not just surviving. Programs like the Harvard Flourishing Project or OECD's well-being indices can be localized using AI to monitor how place-based adaptation affects quality of life across communities.

Advocacy Brief

Issue
Climate change impacts are intensifying at the local level—flooding neighborhoods, overheating cities, and straining local ecosystems. Yet many adaptation efforts remain too generalized to address the specific vulnerabilities of people and places. Communities need solutions tailored to their geographic, cultural, and ecological realities. Artificial Intelligence (AI), when ethically applied, can enhance place-based climate adaptation by offering predictive tools, real-time risk analysis, and community-specific planning support.

Why It Matters
Communities of color, low-income neighborhoods, Indigenous nations, and rural regions often face cumulative environmental impacts—overlapping heat, pollution, poor infrastructure, and displacement. AI can integrate vast data sets—climate projections, land use, health outcomes, traditional ecological knowledge—to identify these compounded risks. When guided by local voices, AI helps ensure that adaptation policies are just, accurate, and rooted in lived experience.

Policy Recommendations

1. Invest in Community-Governed AI platforms that support climate resilience mapping, flood prediction, and heat risk alerts tailored to local needs.
2. Require cumulative impact assessments in all climate adaptation planning, supported by AI-powered data integration.
3. Ensure data equity and sovereignty, especially for Indigenous, rural, and historically marginalized communities.
4. Train local governments and grassroots groups in ethical, place-based AI tools to strengthen public participation in adaptation planning.

Call to Action
We urge policymakers, planners, and funders to bridge technology with place. Support AI systems that are accountable to communities, grounded in geography, and committed to climate justice. A resilient future depends on knowing—and protecting—what makes each place unique.

Planner Toolkit

1. Local Climate Risk Mapping
Use downscaled climate data, hazard maps, and AI tools (e.g., UrbanFootprint, IBM Environmental Suite) to identify place-specific risks like flooding, heat, or wildfire.

2. Cumulative Impact Assessment
Overlay environmental hazards with social vulnerability data (e.g., EJScreen, CalEnviroScreen) to identify areas facing multiple, compounding risks.
3. Community Engagement
Use participatory mapping, cultural asset inventories, and co-design workshops to involve residents in shaping adaptation goals rooted in local identity and needs.
4. Nature-Based and Cultural Solutions
Restore wetlands, forests, or traditional practices (e.g., Indigenous fire stewardship) that support ecosystem health and community resilience.
5. AI and Real-Time Monitoring
Use AI for predictive modeling, early warning systems, and risk surveillance tailored to the geography and demographics of each place.
6. Policy and Zoning Reform
Update land use codes to reflect climate realities—e.g., flood-adaptive housing zones, green buffer areas, or wildfire-resilient development.
7. Equitable Funding and Governance
Access climate justice grants (e.g., Justice40, FEMA BRIC) and ensure frontline communities co-lead planning and resource allocation.
8. Success Metrics
Track outcomes using local well-being indicators and flourishing frameworks, not just infrastructure metrics.

Resources

1. ICLEI – Local Governments for Sustainability. (2024). *Pathways to Resilient Communities: Localizing Climate Action.*
A comprehensive toolkit for local governments implementing place-based adaptation and visioning strategies.
2, UNDP. (2024). *Human Development Report: Building Resilience in Place.* United Nations Development Programme.
Analyzes climate vulnerability in relation to cultural geography, land tenure, and Indigenous knowledge systems.
https://hdr.undp.org
3. Ford, J. D., Cameron, L., & Smith, T. (2024). Reframing Climate Adaptation Through Indigenous Place-Based Knowledge. *Climate & Society, 16(1),* 33–50.
Focuses on Arctic and North American Indigenous approaches to climate adaptation rooted in land-based practices.
4. Doughnut Economics Action Lab. (2025). *City Portraits and Place-Based Sustainability Metrics.*
Provides a framework for local adaptation that balances ecological limits and social foundations using spatial tools.
https://doughnuteconomics.org

5. Cruz, M. J., & Kumar, S. (2024). *Artificial Intelligence in Place-Based Climate Adaptation. Climate Intelligence Review, 11(2), 45–61.*
Examines AI's role in customizing adaptation responses to local climate data, hazards, and community-defined priorities.

6. Natural Resources Canada. (2024). *Community Climate Adaptation: Tools and Stories from Northern Places.*
Features permafrost monitoring, Indigenous fire stewardship, and AI-mapped ecological risk across Canadian bioregions.

7. U.S. Environmental Protection Agency. (2024). *Cumulative Impacts: A Guide for Place-Based Environmental Decision-Making.*
Outlines how agencies can integrate cumulative impact assessments into local adaptation and land use planning.

Chapter Nine
Climate Education Supporting Learning, Adaptation, and Flourishing

"In the end we will conserve only what we love; we will love only what we understand and we will understand only what we are taught."
—*Baba Dioum, Senegalese forestry professor*

A. The Education of the Climate Adaptation Generation

Much of the climate adaptation generation is in school. Approximately 74 million youths under 18 reside in the US, and another 19 million are enrolled in higher education. Not only is this nearly a quarter of the total US population, but it is also the demographic cohort for which we need to create knowledgeable leaders on climate change. This cohort faces the most climate anxiety because it is the cohort that is beginning to see the early impacts of climate change and knows there may be unforeseen climate impacts that will affect their lives.

In Climate Change in the Classroom: Celebrating Optimism for Students, Teachers, and Parents with Multicultural Interdisciplinary Activities, I emphasize that education must evolve to meet the demands of a rapidly changing climate. Climate literacy today is not simply a cognitive process; it is emotional, civic, and deeply interdisciplinary. Across the world, emerging educational resources support this shift, with artificial intelligence (AI) playing an increasingly vital role in transforming how climate adaptation is taught, visualized, and implemented.

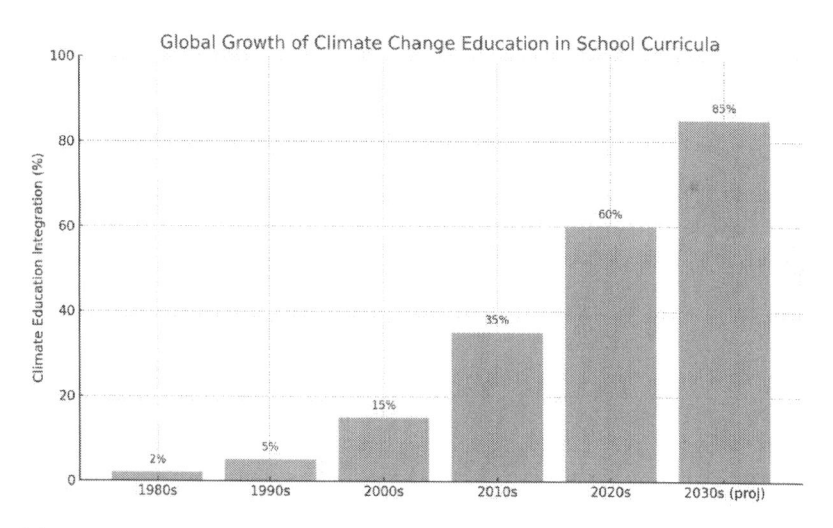

Global Growth of Climate Change Education in School Curricula

Decade	Climate Education Integration (%)
1980s	2%
1990s	5%
2000s	15%
2010s	35%
2020s	60%
2030s (proj)	85%

Traditional Disciplines and Changing Standards

The Next Generation Science Standards (NGSS) were formed in 2013. These standards now apply to at least 20 states. These standards are explicitly designed to provide teachers with the flexibility to adapt lessons. Global Climate Change is now a disciplinary core idea. The target student audience for this focus is primarily middle and high school students, although state approaches may vary. Other core climate change disciplinary ideas focus on primary schools, such as the Human Impacts on Earth's Systems.

The realities of climate change continue to strain all these broader disciplinary boundaries. Teachers are still expected to adhere to a traditional core curriculum composed of individual disciplines. They are still being encouraged to adopt an interdisciplinary teaching approach to engage in climate change education. No single discipline can create successful climate adaptation strategies. That does not require every teacher to be an expert in every discipline. Still, it does require them to be supported by an institutional commitment that includes interdisciplinary collaboration, community engagement, a broad range of inclusivity in climate adaptation interests, and multicultural, global engagement with successful climate adaptation cases.

Multiculturalism, Diversity, and Indigenous Peoples

Climate change is a global issue, and we live in diverse cultures. To develop knowledgeable leaders on climate change, we must instill the ability to understand and engage with other cultures. The text features over 20 successful case studies from diverse cultures, with many more available in the annotated Resources links. Indigenous communities suffer the most from climate change and are the most vulnerable to future climate changes. They have adapted to climate change under many circumstances. The lived experience of successful climate change case studies is especially valuable for fostering optimism, promoting understanding, and alleviating climate change anxiety. That is why multicultural teaching approaches and resources are emphasized throughout the book.

Core Curriculum

Climate education is most effective when it is integrated into core curricula. Tools like NASA's Climate Kids and NOAA's Data in the Classroom enable students to explore climate data in real time and compare local conditions with global trends (NASA, 2025; NOAA, 2024). National Geographic Education continues to provide interdisciplinary connections between climate change, history, cultures, and ecosystems. AI integration in these tools allows real-time pattern recognition, climate risk simulations, and personalized environmental forecasts, making abstract concepts more tangible and locally relevant (Allied Market Research, 2024).

Experiential learning is another cornerstone. Programs like Eco-Schools USA, Project Learning Tree's GreenSchools, and Jane Goodall's Roots & Shoots provide platforms where students conduct environmental audits and lead schoolwide climate actions.

These programs are now enhanced with AI-supported dashboards and geospatial tools, allowing youth to track carbon savings and visualize the outcomes of their adaptation projects (IFRC, 2025; FEMA AI Futures Lab, 2025).

Youth climate leadership is flourishing. Initiatives such as Our Climate, the Sunrise Movement, and the Alliance for Climate Education train students in advocacy, policy communication, and storytelling. AI supports these campaigns by mapping community risks, modeling legislative impacts, and even co-generating policy briefs with students (Cruz & Kumar, 2024).

Climate justice and Indigenous knowledge are critical pillars of contemporary curriculum development. The Indigenous Environmental Network and UNESCO's global Indigenous Knowledge platform offer valuable resources rooted in traditional ecological knowledge. AI tools can now assist in preserving oral histories, mapping ancestral lands, and modeling the disproportionate effects of climate change on Indigenous communities (International Bioregional AI Partnership, 2025).

B. Climate Anxiety

Addressing climate anxiety is essential for climate change adaptation. There are good reasons to feel anxious. Public health is increasingly threatened by climate change. Educational platforms like Climate for Health, Harvard's Center for Climate, Health, and the Global Environment (C-CHANGE), and the Medical Society Consortium provide classroom tools that highlight the connection between air pollution, heat, water safety, and human health. AI assists with public health mapping and early warning systems, particularly in low-income communities facing cumulative environmental burdens (Munich Climate Risk Lab, 2024).

The Climate Mental Health Network and the Good Grief Network offer school-based tools for building emotional literacy, providing safe spaces for expression, and connecting students to support systems. AI-enabled apps for educators and youth can track emotional patterns, offer mindfulness exercises, and connect users with mental health professionals when needed (Harvard C-CHANGE, 2024; Climate Mental Health Network, 2024).

C. Confrontational Discussions in the Classroom: Balanced Messaging

When introducing climate change in the classroom, it's pivotal to frame the conversation in a way that empowers students rather than paralyzing them with fear. The fear of the future of escalating, controversial, and unknown climate change impacts can lead to depression. One practical approach is to center the curriculum around solutions-oriented education. This approach can transform initial anxiety and helplessness into proactive engagement and action. Instead of focusing solely on the overwhelming and often distressing nature of climate issues such as floods and wildfires, educators can redirect the narrative to highlight opportunities for positive change that are within reach. Imagine adaptable lesson plans bursting with creativity, such as designing and building solar ovens or participating in hands-on experiences like planting and maintaining school gardens. Such activities provide practical knowledge about renewable energy sources and sustainable living and grant students a sense of agency. These projects serve as gateways to understanding that climate change can involve both global actions and everyday steps that everyone, regardless of age or resources, can take. It's about converting abstract and intimidating concepts into tangible and actionable steps that students can relate to and act on. This goes a long way to reducing climate anxiety and fostering optimism.

Balanced messaging is another cornerstone of teaching climate change without inducing fear. It plays a crucial role in fostering and maintaining a sense of optimism amid the often-disheartening data. This involves presenting climate data and projections alongside narratives of advancements and innovation, conveying hope and motivation. For example, while it is essential to acknowledge the issue of rising sea levels, it is equally vital to introduce students to innovative methods of coastal protection that have been successfully implemented worldwide. Techniques such as natural barrier restoration and constructing green shorelines are becoming increasingly viable and effective.

D. AI in the Classroom

In urban design, resources like Green Schoolyards America and the Urban Drawdown Initiative promote youth participation in community redesign. Students can propose green corridors, rooftop gardens, or carbon sinks, and AI tools now simulate the impact of those interventions under different emissions scenarios (Google Crisis AI Report, 2024).

Storytelling and media literacy tools such as the Story of Stuff Project, PBS LearningMedia, and Climate Generation's Youth Climate Storytelling Toolkit empower students to share their climate journeys. AI platforms allow students to co-create narratives, generate visuals, or simulate environmental futures, deepening engagement and empathy.

Community science is thriving through iNaturalist, NASA's GLOBE Observer, and CoCoRaHS. These platforms train students to collect and upload climate data, making them contributors to large-scale environmental datasets. AI now aids in pattern recognition, predicts species migration, and enhances data quality for researchers and young scientists alike (Natural Resources Canada, 2024).

Place-based and bioregional education grounds students in their local environments. The Children & Nature Network, EcoRise, and the Bioregional Learning Centre provide frameworks for ecological observation and civic action. AI supports this approach by creating hyperlocal climate impact models and mapping water flow, soil changes, or flood zones within specific watersheds (Cruz & Kumar, 2024).

Educators need support. Tools like SubjectToClimate, the Climate Literacy and Energy Awareness Network (CLEAN), and the Zinn Education Project provide peer-reviewed, standards-aligned content. AI assists teachers by recommending personalized content based on location, student interests, and age group (FEMA AI Futures Lab, 2025).

Mental health integration offers strategies for recognizing climate-related emotions, fostering community, and transforming grief into growth. Classroom activities, such as resilience storytelling, climate art showcases, and eco-gratitude journaling, are simple yet powerful ways to center emotional wellness in climate education. AI-powered emotional support tools—like journaling apps or mood trackers—now complement these efforts.

Sample lesson plans in Climate Change in the Classroom span grades K–12 and engage students in real-world problem solving—from local weather tracking to school climate audits. High schoolers are encouraged to lead "Climate Solutions Fairs" that engage their wider communities. Community activities such as intergenerational climate storytelling circles and neighborhood preparedness events reinforce the local, civic, and hopeful tone of the book.

E. Conclusion

In sum, climate education today is dynamic, interdisciplinary, and emotionally grounded. With the support of AI-enhanced tools and up-to-date climate platforms, educators can help students move from fear to action, from uncertainty to vision. Climate Change in the Classroom advocates for a flourishing-centered education—where every student learns not only how to adapt but how to lead.

Advocacy Brief

Purpose:

To promote the integration of climate adaptation education across K–12 curricula as an essential tool for preparing students to understand, respond to, and thrive in a rapidly changing climate.

Why It Matters:

Climate change is no longer a distant threat—it is a present reality affecting ecosystems, economies, and public health. Children are among the most vulnerable populations. Education systems must evolve to equip students with the knowledge, skills, and resilience necessary to navigate climate-related disruptions, including heatwaves, floods, food insecurity, and mental health impacts.

Policy Recommendation:

Mandate climate adaptation as a core interdisciplinary component in public education, emphasizing local solutions, equity, and empowerment. Curricula should include:

- *Place-based learning tied to students' local environments*
- *Scientific literacy on climate systems, risks, and resilience strategies*
- *Emotional resilience tools to address climate anxiety and promote hope*
- *Civic engagement opportunities to encourage student-led adaptation projects*

Implementation Actions:

- *Update state education standards to include adaptation content*
- *Fund teacher training on climate adaptation and pedagogy*
- *Partner with climate scientists, Indigenous leaders, and planners for curriculum co-design*
- *Embed adaptation into subjects like science, geography, social studies, and the arts*

Call to Action:

Education policymakers, school boards, and local governments must act now to create climate-ready schools that not only teach about climate change but actively prepare the next generation to adapt and flourish.

Toolkit for Planners

Objective:

Empower school districts, urban planners, and education officials to integrate climate adaptation into school design, curriculum, and community planning.

1. Curriculum Integration

- *Embed adaptation topics (heatwaves, droughts, floods, food systems, resilience) across core subjects.*
- *Prioritize place-based and project-based learning tied to local climate risks.*
- *Partner with scientists, Indigenous educators, and local governments for co-developed materials.*

2. School Infrastructure

- *Assess and retrofit schools for heat, flood, and wildfire resilience (e.g., shaded outdoor areas, permeable surfaces, air filtration).*
- *Incorporate green infrastructure: school gardens, rainwater harvesting, bioswales, and solar energy.*

3. Teacher Training

- *Provide professional development on climate literacy and emotional resilience.*
- *Support teachers in using climate data, simulations, and storytelling.*

4. Student Empowerment

- *Support student-led adaptation projects (e.g., cooling maps, emergency kits, community climate audits).*
- *Foster civic engagement through climate councils or local resilience campaigns.*

5. Equity & Justice

- *Prioritize climate adaptation resources for underserved and high-risk school communities.*
- *Include cultural knowledge and justice-centered pedagogy in all adaptation efforts.*

Quick Start Actions

- *Conduct school vulnerability assessments*
- *Integrate adaptation goals into educational master plans*
- *Apply for funding from climate-resilience and green school grants*
- *Engage students and families in planning processes*

Resources

1. Allied Market Research. (2024). *AI for disaster response: Market forecast 2025–2035*. https://www.alliedmarketresearch.com
2. Climate Mental Health Network. (2024). *Resources and toolkits for climate educators*. https://www.climatementalhealth.net
3. Cruz, M. J., & Kumar, S. (2024). Artificial intelligence in climate adaptation and disaster resilience. *Climate Intelligence Review, 11*(2), 45–61.
4. FEMA AI Futures Lab. (2025). *Integrating machine learning in federal disaster planning*. US Department of Homeland Security.
5. Google Crisis AI Report. (2024). *Flood Forecasting and AI-Powered Alerts: Lessons from South Asia*. Google Research.
6. Harvard C-CHANGE. (2024). *Climate, Health, and Education Resources*. Center for Climate, Health, and the Global Environment. https://www.hsph.harvard.edu/c-change
7. IFRC Simulation & Tech Futures Report. (2025). *Next-gen emergency responder training with AI and VR*. International Federation of Red Cross and Red Crescent Societies.
8. International Bioregional AI Partnership. (2025). *Cross-border AI solutions for climate emergencies in shared ecosystems*. Global Climate Response Network.
9. Munich Climate Risk Lab. (2024). *AI and insurance: Redesigning risk in the era of climate volatility*. Munich Re.
10. NASA. (2025). *Climate Kids*. https://climatekids.nasa.gov
11. National Oceanic and Atmospheric Administration (NOAA). (2024). *Data in the Classroom*. https://dataintheclassroom.noaa.gov
12. Natural Resources Canada. (2024). *AI-guided wildfire management in Canada: 2023 pilot*. https://www.nrcan.gc.ca
13. 17 Meaningful Climate Change Activities for Kids. March 26, 2025, Jill Staake. *17 Meaningful Climate Change Activities for Kids*.
14. 20 School Project Ideas to Teach About Climate Change, March 4, 2025. *20 School Project Ideas to Teach About Climate Change*.
15. The New Toolkit Opens Path for Teachers to Start Critical Conversations in Schools: Easy, Short, and Effective. April 27, 2025. Tina Deines. *A new toolkit opens a path for teachers to start critical school conversations: 'Easy, short, and effective.'*
16. The En-ROADS Climate Workshop https://www.climateinteractive.org/the-en-roads-climate-workshop. https://greenerinsights.com/20-school-sustainability-project-ideas-to-engage-students/. March 4, 2025. Lee Hill.

17. Climate Storytelling: A Quick Guide. 2025. APHA Center for Climate, Health and Equity. https://www.apha.org/-/media/Files/PDF/topics/climate/Climate_Storytelling_Guide.pdf.
18. Growing climate-conscious children: early interventions. 2025. https://firstyearsfirstpriority.eu/growing-climate-conscious-children-early-interventions-for-climate-literacy/.
19. Hands-On Activities for Teaching Climate Change: Engaging Strategies for the Classroom. 2025. Marise Sorial, Educator Review Michelle Connolly. Hands-On Activities for Teaching Climate Change: Engaging Strategies for the Classroom - LearningMole.
20. 5 Interactive Climate Change Education Tools to Wow Your Students. February 9, 2023. Fiona Jones, editor.
21. Climate Reanalyzer. By the Climate Change Institute and the University of Maine. 2025.
22. 5 Creative Ways to Teach Human Impact on Climate Change. 2025. Labster. https://www.labster.com/blog/5-creative-ways-teach-human-impact-climate-change.

Chapter Ten

Climate Adaptation and the Health of the Climate Adaptation Generation

Dr. Maria Neira, Director of Public Health, WHO

"The climate crisis is a health crisis. We are witnessing more cases of vector-borne diseases in new areas, heatwaves killing thousands, and air pollution damaging lungs and brains—especially in children. Health systems are being pushed to their limits."
— *Interview with WHO Newsroom, March 2025*

A. The Climate Change Generation and the Risks They Face

The climate adaptation generation—today's youth born into a world already shaped by climate disruption—faces a growing set of public health challenges that are both immediate and long-term and often deeply interconnected. Unlike previous generations, their formative years are defined not by the distant threat of climate change but by its active and accelerating presence in their daily lives. [GU1]These public health impacts fall into three interwoven categories: direct health threats, indirect systemic disruptions, and synergistic effects that magnify vulnerabilities.

Direct health impacts are those resulting immediately from climate-related environmental changes. Intensifying heatwaves have become a defining feature of summers worldwide, with young people particularly vulnerable to heat exhaustion, heatstroke, and dehydration due to their still-developing thermoregulation systems. Children attending schools without air conditioning, particularly in underfunded districts, face severe health risks simply by sitting through a school day. The rising frequency and severity of extreme weather events—including hurricanes, wildfires, floods, and prolonged droughts, introduce widespread injury, trauma, and displacement. These disasters can have lasting impacts on physical health, including burns, respiratory illnesses caused by wildfire smoke, or infections from contaminated floodwaters. Furthermore, changes in temperature and precipitation patterns have expanded the habitat ranges of disease-carrying insects, such as mosquitoes and ticks, thereby increasing the risk of vector-borne diseases, including dengue fever, malaria, Lyme disease, and Zika virus. For younger populations with developing immune systems, these diseases can be particularly debilitating or fatal.

Beyond immediate health consequences, the climate adaptation generation must also contend with a wide array of indirect effects that emerge through disruptions to food, water, shelter, and ecosystem stability. Food security is increasingly compromised by crop failures, declining yields, and lower nutrient content in staple crops such as wheat and rice—an effect linked to rising atmospheric CO_2 concentrations. As a result, children are facing increased risks of undernutrition, stunting, and related developmental disorders, especially in areas where subsistence agriculture is a primary food source. Water scarcity is another major concern, as droughts reduce access to clean drinking water while floods overwhelm sanitation infrastructure. This contributes to higher rates of diarrhea, waterborne illnesses, and long-term exposure to chemical contaminants such as nitrates and arsenic, which have been linked to cancers and neurological disorders. The compounding effects of these environmental stressors are already straining the resilience of public health systems worldwide.

Perhaps one of the most insidious impacts is the rise of mental health burdens. The psychological toll of living through ongoing climate disasters, combined with bleak forecasts of environmental decline, has led to an unprecedented increase in eco-anxiety, grief, depression, and trauma among young people. Many report feelings of hopelessness and existential dread, fearing that they will inherit a damaged planet with fewer options for a safe and meaningful life. Studies show that repeated exposure to climate-related disasters—such as losing a home to fire or flood—can result in post-traumatic stress disorder (PTSD), especially when recovery systems are slow or inequitable. For youth who are displaced by climate-induced migration, the sense of identity loss, cultural disruption, and educational instability can further compound emotional and psychological strain.

B. Urbanization

These public health challenges do not exist in isolation; they interact in complex and often synergistic ways. Urban areas, for instance, experience a combination of heat island effects and poor air quality, especially in neighborhoods lacking green space. This combination intensifies respiratory problems such as asthma and contributes to cardiovascular disease, disproportionately affecting children in low-income communities.

As of 2023, more than half—approximately 57–58%—of the world's population resides in urban areas. This marks a dramatic shift from 1950 when only about 30% of the population lived in cities. Urbanization reached a milestone in 2007 when the urban population surpassed the rural population for the first time. By 2018, cities housed roughly 55%, increasing to 56–57% by 2023. Looking ahead, the UN projects approximately 68% of people will live in urban areas by 2050—with total urban dwellers rising from around 4.0 billion today to nearly 7 billion. This wave of urban growth is expected to be concentrated in Asia and Africa, accounting for 90% of the increase. Two-thirds of the global population is projected to live in urban areas by 2050, driven by population growth in Asia and Africa.

Regional Trends

- High-income regions (e.g., Western Europe, North America, Australia, Japan) already have urbanization rates exceeding 80%.
- Upper-middle-income countries (e.g., those in Eastern Europe, South America, North Africa, and East Asia) typically fall between 50% and 80%.
- Low-income regions, particularly Sub-Saharan Africa and parts of Asia, have lower urban shares—around 40%–50%—but are experiencing the fastest urban growth rates.

Tall buildings can create an urban canyon effect that blocks wind flow, which would otherwise provide ventilation and cool the streets below, as well as speed up evaporation. Tall buildings can also block the release of heat energy into the atmosphere, keeping it closer to the Earth's surface. This traps more heat energy than humans can feel.

C. Waste Heat

Densely populated urban areas concentrate heat-emitting devices, such as cars and air conditioners, over a small area. All this heat adds up and contributes to higher air temperatures in cities.

Over 80% of Americans live in urban areas, and the Urban Heat Island effect means that those metropolitan areas are likely hotter than rural areas. Because of this, urban areas tend to concentrate exposure and experience higher temperatures. The Centers for Disease Control and Prevention (CDC) estimates that the annual rate of heat-related deaths per 100,000 population is 0.3 in large central metro areas –the highest mortality rate paralleled only by "noncore" rural areas, which also experience a mortality rate of 0.3.

In addition to mortality, higher temperatures compromise human health and comfort, with an increased risk of respiratory illnesses, heat exhaustion, heat stroke, and heat-related mortality. They also increase the energy consumption required for air-conditioning homes and buildings, leading to higher emissions of air pollutants and heat-trapping gases, which exacerbate climate change.

Urban heat is not distributed evenly across a city. Neighborhoods in the same city at the same time can differ in temperature by 20°F, primarily due to variations in the factors that contribute to the urban heat island effect.

D. What Can Be Done to Address Urban Heat?

Community leaders and individuals can take many actions to reduce urban heat. Many of these actions have additional co-benefits:

- Plant trees along streets to make shade, especially over dark surfaces (learn about the co-benefits of urban forestry at the Vibrant Cities Lab)
- Add vegetation to urban spaces, including green roofs.
- Implement cool surfaces on roofs, roads, and walls (learn more about cool surfaces from the Global Cool Cities Alliance)
- Provide more access to public air-conditioned spaces.
- Vary the height of new buildings to increase airflow and create shade canyons.
- Use more natural ventilation in buildings.

- Take advantage of programs like <u>LIHEAP</u> and <u>WAP</u> to make energy more affordable and reduce waste by maintaining a safe temperature at home.
- Additional options for managing urban heat are available in the American Planning Association's report, "<u>Planning for Urban Heat Resilience</u>."

In some regions, the convergence of high temperatures, disease outbreaks, and food shortages leads to cascading health risks—malnutrition weakens immune systems, increasing susceptibility to illness. At the same time, concurrent heat waves amplify dehydration and organ stress. In areas experiencing political instability or conflict exacerbated by climate stressors, public health infrastructure is further undermined, leaving young people without consistent access to healthcare, education, or psychological support.

E. The Most Vulnerable Are the First

The most vulnerable within the climate adaptation generation—infants, disabled youth, Indigenous children, and those in low-income or displaced communities—face amplified risks. Infants are particularly susceptible due to their underdeveloped immune systems and limited ability to regulate body temperature. Youth with disabilities may lack access to emergency response systems, evacuation routes, or inclusive adaptation planning. Indigenous and rural children often experience cultural and ecological loss as their traditional lands degrade or are rendered uninhabitable, undermining mental health and identity.

F. Cumulative Impacts

The cumulative effect of these impacts is straining global health systems. Hospitals, clinics, and supply chains are frequently unprepared for the frequency and scale of climate-related emergencies. Healthcare workers, including pediatricians, school nurses, and emergency responders, are increasingly reporting burnout due to repeated cycles of crisis. Public health infrastructure, particularly in the Global South and underserved urban regions—is not always resilient to climate extremes, leading to gaps in care at precisely the moments when it is most needed.

The climate adaptation generation is navigating a world where climate change is no longer a distant threat but an everyday reality that shapes their physical, mental, and developmental health. Their well-being is being reshaped by ecological instability, and their future depends on how effectively public health systems adapt.

Addressing these risks requires comprehensive strategies that combine disaster preparedness, mental health services, climate-resilient healthcare infrastructure, and intergenerational equity. Investing in this generation's health is not only a moral imperative—it is foundational to humanity's ability to flourish in a changing climate.

G. Disease Expansion

The expansion of disease vectors due to climate change is one of the most pressing and complex public health threats facing the current generation of climate adaptors. As global temperatures rise, precipitation patterns shift and ecosystems are altered, the habitats and life cycles of disease-carrying organisms, such as mosquitoes, ticks, fleas, and rodents, are undergoing dramatic changes. This has profound implications for the spread of infectious diseases, many of which were once geographically limited but are now appearing in new regions, affecting populations with little or no immunity.

The rise of vector-borne diseases fueled by climate change is having significant and often deadly consequences for populations around the world, especially among children, pregnant women, and those in low-resource settings. As global temperatures climb and precipitation patterns shift, disease vectors such as mosquitoes, ticks, and rodents are spreading into new geographic areas, elevating the risk of exposure to previously localized illnesses.

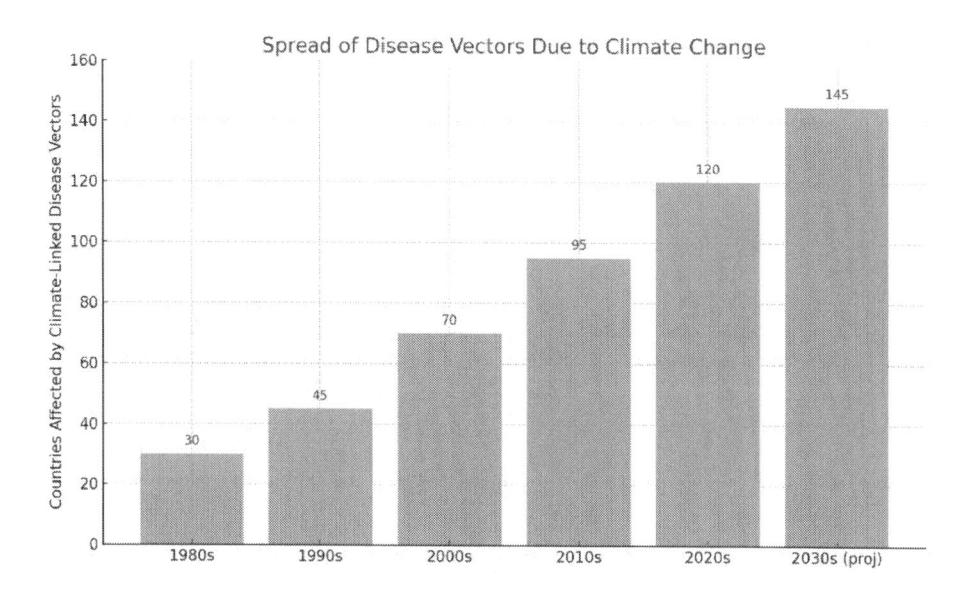

H. The Diseases

Dengue fever, carried by Aedes aegypti and Aedes albopictus mosquitoes, is one of the fastest-growing global threats. Infections begin suddenly, with fevers reaching up to 104 degrees Fahrenheit, accompanied by intense headaches, pain behind the eyes, severe muscle and joint aches that have earned the nickname "breakbone fever," and a characteristic skin rash. Nausea, vomiting, and fatigue are also common. While most cases resolve within a week, severe dengue, also known as dengue hemorrhagic fever, can be fatal if not treated promptly. This form causes internal bleeding, plasma leakage, shock, and organ failure. Without medical care, mortality in severe cases can reach up to 20%, though with proper treatment, it drops below 1%.

Malaria, caused by Plasmodium parasites and spread by Anopheles mosquitoes, remains a major killer, especially in sub-Saharan Africa. Symptoms typically begin with cyclical fevers, chills, muscle aches, headaches, and nausea. If left untreated, malaria can progress rapidly, particularly the Plasmodium falciparum strain, leading to cerebral complications, respiratory failure, anemia, and death —sometimes within just 48 hours. Young children are especially vulnerable, and malaria remains responsible for the deaths of over 600,000 people annually. The mortality rate can soar to 15% in untreated severe pediatric cases, though it remains below 1% when effective treatment is promptly administered.

Zika virus, another disease transmitted by Aedes aegypti, typically causes relatively mild symptoms in most infected individuals. These may include fever, rash, red eyes, joint pain, and muscle soreness, usually lasting less than a week. However, the danger lies in its impact on pregnant women and unborn children. Zika is directly linked to congenital Zika syndrome, which causes microcephaly —abnormal brain and skull development—and other severe congenital disabilities. Though the virus rarely causes death in adults, the long-term disabilities it inflicts on infants have lifelong consequences for families and public health systems.

Chikungunya, also transmitted by Aedes mosquitoes, causes a sudden onset of high fever and excruciating joint pain. Patients often suffer from headaches, rashes, and muscle inflammation. While the disease rarely results in death, it leaves many with chronic arthritis-like symptoms that may last for months or even years. The mortality rate is very low—less than 0.1% in most populations— but can rise to 1–2% among elderly or immunocompromised individuals. Though not deadly for most, the long-term disability burden from Chikungunya is significant and economically disruptive.

In temperate regions such as North America and Europe, Lyme disease has become increasingly common due to the spread of Ixodes ticks. Early signs include a distinctive bull's-eye rash, fever, fatigue, headaches, and muscle aches. If untreated, the disease can progress to more serious complications such as arthritis, neurological impairment (including facial paralysis and memory loss), and cardiac arrhythmias. While Lyme disease is rarely fatal, it can result in long-term disability when diagnosis is delayed, or treatment is inadequate.

Tick-borne encephalitis (TBE), which is also on the rise in parts of Scandinavia and Central Europe, begins with flu-like symptoms such as fever and fatigue but can progress to meningitis and encephalitis. When the central nervous system is involved, patients may experience seizures, confusion, speech disorders, and paralysis. Even with medical care, the mortality rate ranges from 1 to 2%, with up to one-third of survivors suffering lasting neurological damage.

West Nile virus, carried by Culex mosquitoes, typically causes no symptoms in most individuals. However, in some cases, people experience fever, headaches, vomiting, body aches, and rashes. In about one in 150 cases, the virus invades the nervous system, leading to encephalitis, meningitis, paralysis, or even coma. Neuroinvasive cases have a mortality rate of around 10%, with elderly individuals at significantly higher risk of death or permanent damage.

In flood-prone cities like Mumbai, Dhaka, and parts of the Caribbean, leptospirosis has become a recurring post-disaster health threat. Spread through water contaminated with the urine of infected rodents, this disease causes fever, red eyes, vomiting, and muscle pain. When it develops into Weil's disease, it can lead to kidney failure, liver damage, and pulmonary hemorrhage. The severe form carries a mortality rate as high as 15%, and in cases involving lung complications, mortality can exceed 40% if not treated intensively.

Another rodent-borne illness exacerbated by environmental shifts is hantavirus pulmonary syndrome (HPS). Found mainly in the Americas, it causes flu-like symptoms that escalate into acute respiratory failure, often within a few days. There is no specific treatment or cure, and the mortality rate ranges from 30 to 40%, making it one of the deadliest climate-sensitive diseases currently known.

These diseases, while biologically distinct, share common vulnerabilities: they thrive under the ecological conditions intensified by climate change, and they expose weaknesses in public health infrastructure, surveillance, and healthcare access. Their symptoms range from mild discomfort to catastrophic disability or death, and their mortality rates are closely tied to how quickly individuals receive appropriate care.

In many regions, particularly those experiencing poverty, conflict, or displacement, delayed diagnosis and treatment contribute to unnecessary suffering and long-term public health burdens.

In the era of climate instability, preventing these diseases requires not only medical readiness but also robust environmental health systems, early warning mechanisms, public education, and equitable access to care. Artificial intelligence helps to address these challenges, but without global cooperation and investment, the human cost will continue to rise—especially for those least responsible for climate disruption.

Dengue fever is now endemic in over 120 countries, including regions where it was previously rare or nonexistent. The World Health Organization (2025) has warned that rising temperatures and increased rainfall are creating ideal breeding conditions for mosquitoes in both urban and rural areas. The Aedes aegypti mosquito, which transmits dengue, thrives in warm, wet environments, and its range is rapidly expanding due to climate instability. By 2050, over six billion people may be at risk of dengue infection, compared to 3.5 billion today, marking one of the most dramatic increases in global disease exposure ever projected.

Malaria, a disease historically confined to tropical and subtropical lowlands, is now appearing at higher altitudes in regions such as East Africa, South America, and parts of Southeast Asia. Climate warming shortens the developmental cycle of Plasmodium parasites within Anopheles mosquitoes, thereby increasing the efficiency of transmission. The highlands of Kenya, Ethiopia, and Uganda— once refugees from malaria—are now experiencing regular outbreaks, threatening communities that lack prior exposure and immunity.

Zika virus outbreaks, closely associated with severe congenital disabilities such as microcephaly, have also expanded due to climate change. The Aedes aegypti mosquito has expanded its range into Latin America, the southern United States, and parts of Europe, introducing the virus to populations where it had previously not been found. These outbreaks have had especially devastating consequences for pregnant women and their babies and have exposed significant gaps in reproductive health systems during public health emergencies.

Similarly, Chikungunya disease, once confined mainly to parts of Africa and Asia, has now spread throughout the Americas and southern Europe. Its expansion is driven by the same climatic shifts: warmer temperatures, heavier seasonal rains, and the proliferation of urban breeding grounds such as stagnant water in discarded containers, rooftop cisterns, and poorly maintained drains. Infected individuals experience intense joint pain, fever, and fatigue, and many suffer long-term effects.

Ticks, particularly Ixodes scapularis (commonly known as the black-legged or deer tick), are also migrating to new regions due to shorter winters and longer warm seasons. This range expansion has led to increased rates of Lyme disease, babesiosis, anaplasmosis, and Powassan virus. Lyme disease, once limited to the Northeastern and Upper Midwestern United States, has surged across Canada and into the Northern Plains. Warmer winters enable ticks to survive the cold season and initiate feeding earlier in the spring, thereby extending their seasonal activity.

Powassan virus, a rare but deadly encephalitic illness, has emerged in the northeastern United States and parts of eastern Canada. Unlike Lyme disease, the Powassan virus can be transmitted within just 15 minutes of a tick bite, making it significantly more challenging to prevent. Across Europe, tick-borne encephalitis (TBE) is also becoming more common in areas such as Germany, Switzerland, and Scandinavia—regions that were once considered too cold for ticks to thrive. Warmer, wetter conditions now support the full lifecycle of these vectors and the pathogens they carry.

Rodent-borne and zoonotic diseases are also evolving in both scope and intensity as climate change alters the behavior and habitats of animals. Hantavirus pulmonary syndrome, primarily carried by deer mice in the Americas, tends to spike after wet seasons increase food availability and rodent populations, followed by dry conditions that drive them into human dwellings. This pattern has been seen increasingly in North and South America.

Leptospirosis, a bacterial disease spread through water contaminated with rodent urine, has become more common in flood-prone cities of Southeast Asia, Central America, and even in parts of the United States. Urban flooding associated with intense storms creates perfect conditions for outbreaks in informal settlements with poor sanitation. In West Africa, the incidence of Lassa fever, an arenavirus carried by Mastomys rats, fluctuates with rainfall and temperature shifts that influence rodent breeding and behavior. Additionally, the Nipah virus, linked to fruit bats, poses a growing concern due to its high mortality rate and ability to transmit between humans. Climate change and habitat loss drive bats closer to human populations, increasing the risk of spillover events that could spark future pandemics.

Waterborne and diarrheal diseases, often linked to climate-driven vector dynamics, are surging in flood-stricken and low-lying areas. Following extreme rainfall events, stagnant water creates breeding grounds for mosquitoes and becomes contaminated with fecal matter, contributing to outbreaks of cholera, giardiasis, cryptosporidiosis, and typhoid fever.

These illnesses disproportionately affect children under five, whose smaller bodies and developing immune systems make them especially vulnerable to dehydration and death. Rising sea surface temperatures also promote the proliferation of Vibrio bacteria, including Vibrio cholerae and Vibrio vulnificus, which thrive in brackish waters. These pathogens are increasingly found in regions once considered too cold for survival, including the Baltic Sea and the U.S. Gulf Coast.

Climate change not only introduces new diseases but also shifts the balance between vector species. In warmer regions, traditional mosquito species are being displaced by more heat-tolerant, aggressive vectors that transmit diseases more efficiently. Climate-driven wildfires and deforestation also force vector species into closer proximity to humans, thereby increasing the likelihood of zoonotic spillovers. El Niño–Southern Oscillation (ENSO) events, intensified by global warming, cause cyclical changes in rainfall and humidity that supercharge breeding cycles and drive periodic outbreaks with increasing intensity and unpredictability.

One of the most significant challenges in managing these emerging threats is the gap between the spread of disease and the capacity for surveillance. Many low- and middle-income countries lack the infrastructure to track and respond to new vector-borne illnesses. Underdiagnosis, misclassification, and delayed response are common. Even in high-income countries, health providers unfamiliar with tropical or re-emerging diseases may overlook early warning signs, contributing to preventable deaths and transmission.

I. AI and Disease: Examples

Across the globe, however, artificial intelligence is emerging as a powerful tool in managing this growing threat. In sub-Saharan Africa, where malaria is migrating to higher altitudes, projects like the Malaria Atlas Project use machine learning to map disease suitability based on rainfall, temperature, and vegetation. These models allow health officials to deliver supplies and interventions preemptively and even predict drug resistance using genomic sequencing of the malaria parasite.

In Southeast Asia, dengue outbreaks are being predicted through national AI systems that analyze urban heat, construction activity, and meteorological data. In cities like Manila and Jakarta, these forecasts now inform hospital readiness and mosquito control campaigns. Vietnam's Ministry of Health has adopted Singapore's AI models, while BlueDot software continuously scans digital signals—from social media to hospital records—to provide real-time alerts.

Latin America has integrated similar AI systems to tackle Chikungunya and Zika. In Brazil and El Salvador, AI dashboards map heat anomalies, congenital disability reports, and vector data to identify high-risk neighborhoods. Drone surveillance in Guatemala even uses computer vision to locate urban mosquito breeding sites in real-time.

In the United States, climate change is pushing Lyme disease and West Nile virus into new regions. AI partnerships between the CDC and Google DeepMind have produced weekly risk forecasts across 17 states. Image-recognition apps now allow citizens to submit photos of ticks and mosquitoes for instant species identification and risk classification, feeding data back into public health systems.

In Scandinavia, where tick-borne encephalitis has expanded northward, the EU's VectorNet AI system tracks shifts in tick habitats and integrates satellite data, deer migration, and citizen reports. In Finland, AI models now forecast tick density and disease transmission risk every week, helping to guide vaccination efforts and inform public warnings.

In flood-prone cities of South Asia, AI models are helping predict post-monsoon outbreaks of leptospirosis. The South Asia Disease Risk Engine uses flood depth, elevation, housing density, and sanitation data to generate targeted alerts. At the same time, AI-powered chatbots educate residents in local languages on symptoms and treatment options.

Taken together, these cases highlight how climate change is reshaping the global landscape of infectious diseases. The climate adaptation generation is growing up in a world where vector-borne illnesses are not only more common but more complex. As disease ranges expand and health systems strain under compounding crises, AI offers vital hope—but only if paired with investments in equity, infrastructure, and education.

J. Conclusion: AI as a Critical Tool in Climate-Health Adaptation

Across regions, artificial intelligence is emerging as a vital component in climate adaptation for controlling vector-borne diseases. By integrating diverse data sources—from satellite imagery and meteorological patterns to real-time social media and medical records, AI enables earlier warnings, targeted interventions, and more efficient public health responses. However, these benefits are not evenly distributed. Low- and middle-income countries face challenges in accessing data, developing digital infrastructure, and enhancing technical capacity.

For the climate adaptation generation to thrive, global investments must ensure equitable access to AI tools, support open-source disease forecasting platforms, and strengthen interdisciplinary collaborations between epidemiologists, climate scientists, and community leaders.

Advocacy Brief

June 2025

As climate change accelerates the spread of infectious diseases—including malaria, dengue, Zika, Lyme disease, and leptospirosis—public health systems must adapt rapidly. Rising temperatures, shifting rainfall, and expanding habitats for mosquitoes, ticks, and rodents are transforming the global disease landscape. Climate adaptation generation is especially at risk.

Artificial Intelligence (AI) offers a vital opportunity to protect lives and strengthen climate-health resilience. AI tools are now being utilized to forecast outbreaks, map the migration of disease vectors, automate diagnosis, and provide early warnings to at-risk communities.

Key Public Health Benefits of AI

- *Predictive Accuracy: AI can forecast outbreaks of dengue, malaria, and West Nile virus weeks in advance using satellite, weather, and mobility data.*
- *Faster Surveillance: AI platforms like BlueDot and HealthMap detect new disease clusters from digital signals—days before official alerts are issued.*
- *Risk Mapping: Machine learning models track climate-driven changes in the ranges of mosquitoes and ticks, enabling the targeting of vector control and vaccination efforts.*
- *Equity in Access: AI-powered mHealth tools and chatbots enhance diagnosis and education in low-resource areas, particularly in the aftermath of disasters.*

Urgent Actions Needed

1. *Invest in open-source, climate-sensitive AI tools for disease forecasting, particularly in regions of the Global South.*
2. *Integrate AI into national and municipal health planning, including emergency preparedness and climate adaptation strategies, to enhance overall health outcomes.*
3. *Build digital and health infrastructure in vulnerable communities to ensure equitable access to the benefits of AI.*
4. *Foster ethical AI governance, ensuring transparency, local data control, and community participation.*

AI is not a luxury—it is a frontline defense against climate-induced disease and death. With timely investment and inclusive deployment, it can save millions of lives.

For more information or collaboration, contact:
Global Health and Climate Adaptation Taskforce
Email: info@climatehealthai.org | Web: www.climatehealthai.org

Planners Tool Kit

Planning Tools for Climate-Driven Disease Adaptation

1. Early Warning and Disease Forecasting Platforms

Planners need access to real-time disease forecasting to allocate resources and prevent outbreaks.

- *Malaria Atlas Project (MAP)*
- *Provides high-resolution malaria risk maps based on AI-driven climate and vector data.*
- *https://malariaatlas.org*
- *BlueDot*
- *AI platform that uses air travel, climate, and digital data to predict emerging disease hotspots.*
- *https://bluedot.global*
- *HealthMap (Boston Children's Hospital)*
- *Aggregates outbreak data from news, social media, and official reports into global surveillance maps.*
- *https://www.healthmap.org*

2. Climate-Disease Risk Mapping and Decision Support

Spatial tools help identify where climate impacts will intersect with vulnerable populations.

- *CDC's Environmental Public Health Tracking Network*
- *Interactive maps linking climate indicators (e.g., heat, floods) with disease trends in U.S. regions.*
- *https://ephtracking.cdc.gov*
- *WHO Climate and Health Country Profiles*
- *Provides national-level projections of health risks, including vector-borne diseases, under various climate scenarios.*
- *https://www.who.int/activities/building-capacity-on-climate-change-and-health Climate Adaptation Explorer (Florida) Region-specific tool showing how climate factors affect diseases like dengue, heat stress, and water quality. https://floridaclimateinstitute.org/tools*

3. Urban Health Resilience and Infrastructure Planning

Cities require integrated public health and land-use strategies for effective disease prevention.

- *ICLEI – Climate-Resilient Health Systems Framework*
- *Guides integrating health surveillance into local climate adaptation planning.*
- *https://iclei.org*
- *U.S. Climate Resilience Toolkit – Human Health Sector*
- *Provides resources for local planners to prepare for diseases exacerbated by heat, flooding, and vector-borne diseases.*
- *https://toolkit.climate.gov*
- *UN-Habitat's Urban Health Equity Assessment and Response Tool (Urban HEART)*
- *It helps identify where urban inequality, infrastructure, and climate risks converge, thereby elevating disease burdens.*
- *https://unhabitat.org*

4. Digital and AI Tools for Disease Surveillance and Intervention

Machine learning, mobile platforms, and AI enhance real-time decision-making.

- *PAHO/WHO Dengue AI Lab*
- *Combines congenital disability data, temperature anomalies, and mosquito density to target dengue and Zika responses in Latin America.*
- *https://www.paho.org*
- *Tracks real-time evolution of pathogens (like dengue or Zika) using open-source genomic surveillance data.*
- *https://nextstrain.org*
- *AI4Ticks and VectorNet (Europe)*
- *Predictive modeling tools that integrate environmental and biodiversity data to forecast tick-borne disease risks.*
- *https://www.ecdc.europa.eu/en/about-us/partnerships/vectornet*

5. Community-Based and Health System Planning Resources

Planners must also coordinate with public health agencies and local communities to ensure readiness and preparedness.

- *Community Assessment for Public Health Emergency Response (CASPER)*
- *A CDC methodology for rapidly gathering community health needs during climate-related events (e.g., floods, vector outbreaks).*
- *https://www.cdc.gov/nceh/casper*
- *Climate Resilient Health Systems (CRHS) Toolkit – WHO*
- *https://www.who.int/publications/i/item/9789240068349*
- *Local Environmental Health Planning Guides (NACCHO)*

Summary for Planners

To effectively reduce disease burdens linked to climate change, planners should:

- Integrate vector and disease surveillance into land use and housing decisions.
- Utilize AI and GIS tools to predict outbreaks and allocate medical resources effectively.
- Design green infrastructure and drainage to reduce mosquito habitats in urban areas.
- Promote health equity by targeting interventions for vulnerable populations.
- Collaborate with public health agencies to ensure that climate adaptation includes
- infectious disease preparedness.

Resources

1. World Health Organization. (2025). *Global Vector-Borne Disease Outlook.*
2. Centers for Disease Control and Prevention (CDC). (2024). *Climate Change and Vector Expansion: Annual Report.*
3. IPCC. (2023). *Sixth Assessment Report: Health Impacts from Climate Change.*
4. Lancet Countdown. (2024). *Health and Climate Change: Tracking Progress on Vector-Borne Disease.*
5. National Institute of Environmental Health Sciences (NIEHS). (2024). *Climate Change and Infectious Disease Risks.*
6. UNICEF. (2024). *Climate Crisis and Child Health. Infectious Disease Threats.* Allied Market Research. (2024). *AI for disaster response: Market forecast 2025–2035.* https://www.alliedmarketresearch.com
7. BlueDot. (2025). *AI-enabled outbreak detection platform: Real-time disease intelligence across borders.* https://bluedot.global
8. Centers for Disease Control and Prevention (CDC). (2025). *AI and vector-borne disease modeling initiative: Technical summary.* U.S. Department of Health and Human Services.
9. Cruz, M. J., & Kumar, S. (2024). Artificial intelligence in climate adaptation and infectious disease resilience. *Climate Intelligence Review, 11*(2), 45–61.
10. European Centre for Disease Prevention and Control (ECDC). (2024). *VectorNet AI surveillance system annual report: Tick- and mosquito-borne diseases in Europe.* https://www.ecdc.europa.eu
11. Google DeepMind. (2025). *AI models for U.S. West Nile Virus prediction: A partnership with CDC.* Google Health Research Division.

112. IBM Research & Pan American Health Organization (PAHO). (2024). AI dashboards for dengue and Zika mapping in Latin America: Data-driven disease control. https://www.paho.org
13. International Bioregional AI Partnership. (2025). Cross-border AI systems for climate-emergent disease surveillance in shared ecosystems. Global Climate Response Network.
14. Lancet Countdown. (2024). 2024 report: Tracking the connections between climate change and global health. The Lancet, 404(10380), 1–50. https://www.thelancet.com/countdown-health-climate
15. Nature Medicine. (2024). AI-powered surveillance for emerging vector-borne diseases. Nature Medicine, 30(3), 233–236. https://www.nature.com/articles/d41591-024-00345-2
16. Oxford Malaria Atlas Project. (2024). Machine Learning in Highland Malaria Prediction and Drug Resistance Mapping. University of Oxford, Department of Tropical Medicine. https://malariaatlas.org
17. Singh, P., & Rahman, F. (2025). The rise of AI in climate-sensitive disease control: Ethical, technical, and public health implications. Global Public Health Journal, 18(1), 77–94.
18. UNICEF. (2025). AI and Climate Adaptation: Protecting Children from Disease and Displacement. United Nations Children's Fund. https://www.unicef.org
19. WHO AI and Health Lab. (2024). Artificial intelligence for outbreak prediction and digital epidemiology. World Health Organization. https://www.who.int/initiatives/ai-for-health
20. World Bank. (2024). Leveraging AI for climate-informed health systems in low-income countries: Innovations and barriers. Health, Nutrition, and Population Global Practice Paper.

Chapter Eleven

A Blueprint for Flourishing

" The crisis we face is, first and foremost, one of mind, perception, and values. It is, in short, a crisis of vision. To flourish, we must learn to think in terms of whole systems, long time scales, and shared responsibility."
— Paraphrased from David Orr, Earth in Mind (2004)

The idea of what it means to flourish within a framework of climate adaptation takes a vision.

A. The Relationship Between Visioning and Flourishing in Climate Adaptation

Visioning and flourishing are deeply interconnected in the context of climate adaptation. Visioning is the participatory process by which individuals and communities imagine a desired future grounded in their values, needs, and cultural identity. Flourishing, in turn, is the outcome of this process—a condition of thriving that includes ecological balance, social equity, emotional wellbeing, and intergenerational resilience.

Visioning enables communities to move beyond reactive Adaptation toward proactive transformation. Rather than merely adjusting to climate threats (e.g., heatwaves, floods, wildfires), communities that engage in visioning ask more profound questions: What kind of future do we want to build? What values will guide us? What does it mean to thrive, not just survive, in our place? Through this process, climate adaptation becomes not a burden but a creative and empowering opportunity.

Flourishing expands the goal of Adaptation. Traditional adaptation measures—such as building seawalls or upgrading HVAC systems—often aim to reduce risk. Flourishing insists on more: Adaptation must also improve quality of life, strengthen social bonds, deepen cultural expression, and regenerate the environment. A flourishing community is not just climate-resilient—it is also more just, more connected, and more hopeful.

Both visioning and flourishing rely on collective agency. Visioning centers on local knowledge, intergenerational dialogue, and shared governance. It invites youth, elders, Indigenous peoples, scientists, and artists into the same space to co-create a future. This inclusive process builds trust, ownership, and momentum. In turn, flourishing depends on systems of care, mutual aid, and ecological interdependence, all of which are cultivated through collaborative visioning and planning.

No two communities will flourish in the same way. A coastal village may envision mangrove restoration and tidal farming as potential solutions. An urban neighborhood may focus on green schoolyards, cooling centers, and anti-displacement policies. A rural town may prioritize watershed stewardship and regenerative farming.

Visioning ensures that climate adaptation strategies align with place-specific histories, geographies, and aspirations—key ingredients for true flourishing.

Engaging in visioning enhances climate understanding. By identifying local vulnerabilities, mapping assets, and imagining regenerative futures, participants deepen their systems thinking and civic knowledge. This literacy empowers them to act—and flourishing becomes a self-reinforcing loop: informed communities build resilient systems, and resilient systems sustain conditions for human and ecological flourishing.

In summary, visioning provides the participatory process, and flourishing provides the transformative goal. When integrated into climate adaptation, they shift the focus from emergency response to long-term thriving—rooted in justice, ecology, culture, and care.

A blueprint for flourishing serves as a guiding framework—both comprehensive and adaptable—that helps communities, educators, policymakers, and individuals respond to the realities of climate change while envisioning and co-creating a regenerative future. Like an architect's plan that brings a building into being, a climate blueprint lays out the systems, values, and strategies necessary for shaping communities that are not only resilient but also thriving. It bridges the urgent need for Adaptation with the long-term goal of cultivating lives marked by wellbeing, justice, and ecological harmony.

At the heart of this blueprint is a shared vision rooted in values such as interdependence, sustainability, equity, and lifelong learning. Rather than treating Adaptation as a temporary response to crisis, this model reframes it as an enduring, creative process of transformation. These values shape decisions in education, urban planning, public health, and land use, reinforcing the idea that flourishing must be collective, inclusive, and future-facing.

The process begins with a deep understanding of the local context. Each community faces unique climate-related challenges—such as rising seas, wildfires, and water scarcity—and possesses distinct assets, including natural resources, social networks, and cultural knowledge. A flood-prone region, for example, may focus its blueprint on wetland restoration, adaptive infrastructure, and emergency preparedness. A fire-prone region might prioritize defensible space zoning, Indigenous fire stewardship practices, and heat-resilient housing. The blueprint must be rooted in local realities while connected to global movements for ecological and social justice.

Types of Visioning Under Climate Change

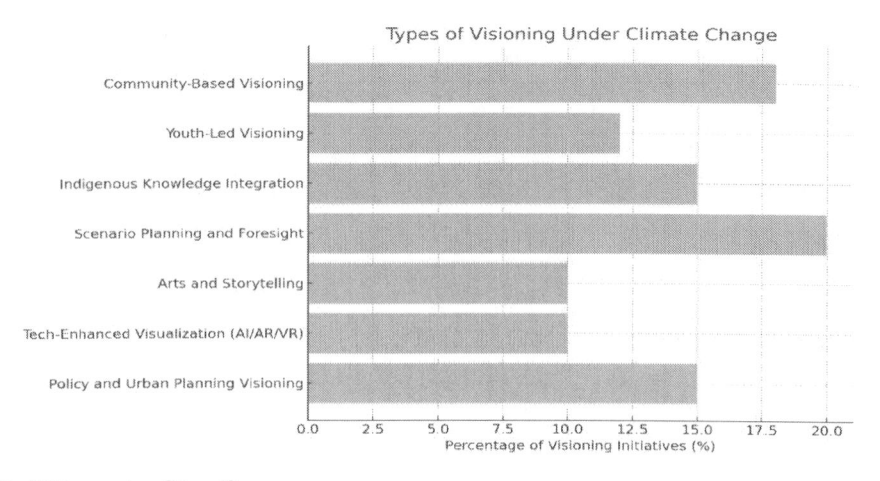

(Bar chart showing percentage of visioning initiatives:
- Community-Based Visioning: ~18%
- Youth-Led Visioning: ~12%
- Indigenous Knowledge Integration: ~15%
- Scenario Planning and Foresight: ~20%
- Arts and Storytelling: ~10%
- Tech-Enhanced Visualization (AI/AR/VR): ~8%
- Policy and Urban Planning Visioning: ~15%)

Percentage of Visioning Initiatives (%)

B. Where to Start?

From this place-based assessment, communities can begin designing systems of Adaptation that are interconnected and multi-benefit. Energy systems transition to solar and wind while strengthening grid resilience. Water systems are restructured to capture, conserve, and reuse resources. Food systems encompass regenerative agriculture, urban gardens, and food forests, which enhance biodiversity and food security. Housing becomes more breathable, insulated, and elevated. Mobility is redesigned for safe, clean, and equitable access. Health services incorporate cooling centers, trauma-informed care, and climate-sensitive medical response. These systems are not isolated—they function as networks that support one another. For instance, Barcelona's "Superblocks" reduce car traffic, improve air quality, and create green public spaces, proving that holistic design fosters both resilience and livability.

However, a blueprint cannot be imposed from above. It must be co-created through democratic, participatory processes. Community members engage through visioning sessions, school workshops, art-based mapping, storytelling circles, and design charrettes. These activities ensure that the blueprint reflects the identities, experiences, and aspirations of the people who will live its outcomes. In New Orleans, for example, intergenerational teams have co-developed neighborhood climate plans that blend local heritage with future resilience, making Adaptation both protective and celebratory.

The next phase involves testing ideas through pilot projects—small-scale, visible efforts that model transformation. A vacant lot is transformed into a climate learning garden. A school installs a rooftop garden and rain catchment system.

A neighborhood hosts a resilience fair. These projects generate feedback and enthusiasm, informing the refinement of the broader plan. Like all living frameworks, a blueprint must evolve as conditions shift, knowledge deepens, and communities grow.

Ultimately, successful blueprints for flourishing become embedded in the institutions and rhythms of public life. They inform zoning codes, guide education standards, and create career pathways in green infrastructure, ecological restoration, and resilience planning. Copenhagen's citywide adaptation strategy has become a model for integrated, cross-departmental planning, resulting in green roofs, flood-absorbing plazas, and citizen involvement as everyday features of urban life.

To sustain momentum, the blueprint includes systems for tracking progress and sharing stories. Metrics such as access to cooling infrastructure, stormwater retention rates, or climate literacy scores help assess impact. But just as importantly, storytelling fosters connection and a sense of belonging. Schools host exhibitions of student climate timelines, elders share ancestral knowledge in public libraries, and artists use murals or performances to express resilience narratives. These acts of storytelling turn data into culture and transformation into celebration.

In summary, a blueprint for flourishing is not a static document. It is a dynamic, inclusive, and visionary approach to climate adaptation that places communities in the role of designers, not just responders. It reminds us that resilience is not only about surviving disruption, it is about growing forward together with care, creativity, and courage.

C. Legislative Frameworks Supporting Visioning and Blueprints for Flourishing

Legislation at local, national, and international levels supports the development of such blueprints by requiring or enabling visioning processes grounded in public participation, sustainability, and long-term planning.

In the United States, comprehensive planning laws in states such as Oregon, California, and Florida require public engagement in community visioning. Oregon's Statewide Planning Goal 1 legally requires local governments to involve citizens in crafting future-oriented land use and climate plans. California's SB 375 mandates Sustainable Communities Strategies that align transportation and housing with climate goals through participatory design.

Florida's Growth Management Act promotes cross-jurisdictional cooperation rooted in long-term visioning.

Federal laws, such as the National Environmental Policy Act (NEPA), embed visioning in the environmental review process. NEPA requires agencies to evaluate multiple future scenarios, incorporating public input, to ensure informed decision-making. The Housing and Community Development Act supports programs like CDBG, which necessitate community-led planning and vision setting for neighborhood revitalization.

In Canada, provincial laws such as Ontario's Planning Act and British Columbia's Local Government Act require municipalities to conduct public visioning in their Official Plans and community strategies. The Pan-Canadian Framework on Clean Growth and Climate Change encourages participatory approaches to climate policy development.

In the European Union, the Strategic Environmental Assessment Directive requires public involvement in major environmental plans, and the Aarhus Convention ensures legal rights to environmental information and public participation—both of which are critical to community-driven planning.

Australia and New Zealand also embed visioning in law. New South Wales requires environmental plans developed through community engagement, while New Zealand's Resource Management Act explicitly incorporates Māori worldviews and intergenerational ecological stewardship into land-use planning.

Indigenous and Tribal governance systems offer some of the most powerful models of visioning. The United Nations Declaration on the Rights of Indigenous Peoples (UNDRIP) affirms the right to Free, Prior, and Informed Consent (FPIC), meaning that any plan affecting their land or people must include their vision. In the United States, tribal nations conduct sovereign planning through Tribal Comprehensive Plans that express culturally rooted futures.

International frameworks further reinforce these principles. Agenda 21 and the Sustainable Development Goals encourage community-based visioning for equitable and sustainable development. The Paris Agreement under the UNFCCC, specifically Article 7 on Adaptation, promotes participatory planning in the development of adaptation strategies.

These legal instruments transform visioning from an optional exercise into a legal and ethical obligation. They ensure that communities have not only the right but also the responsibility to imagine, plan, and build futures that are sustainable and flourishing.

D. Conclusion

Although visioning is a new and vague process for many, it is essential to connect climate change adaptation to flourishing. The reason flourishing is so important is that climate adaptation is not about getting back to the status quo. First of all, the status quo is never possible due to numerous concurrent, unknown, and unpredictable factors, such as ecological changes and global uncertainty. Second, the status quo has produced many unintended impacts detrimental to human health and the environment, which have accumulated to the point of toxicity over time. Last, the status quo's intended impacts were often oppressive and unjust to many people who would not tolerate a return to such practices. New ways of envisioning, such as the goal of flourishing, appeal to the climate adaptation generation, giving them hope, direction, and energy to move forward with climate adaptation plans that benefit everyone.

Advocacy Brief

Purpose

To advocate for the adoption of community-based, legally supported visioning and planning processes—referred to as blueprints for flourishing—that integrate climate adaptation, equity, and regenerative development into public education and local governance.

Background

As the climate crisis intensifies, communities face rising threats, including sea-level rise, extreme heat, droughts, and ecological disruption. Traditional adaptation models often focus narrowly on infrastructure and emergency response. This advocacy brief advocates for a broader, more holistic strategy—embedding climate adaptation into the daily lives of communities through schools, land-use plans, storytelling, and policy. A blueprint for flourishing reframes Adaptation not as survival alone but as a pathway toward equity, creativity, and ecological renewal.

Policy Recommendations

Mandate Visioning in Climate Planning

- *Governments should require public, intergenerational visioning processes in all climate adaptation and land-use strategies. This includes legal mechanisms already present in planning laws such as Oregon's Goal 1 (Citizen Involvement), California's SB 375, and New Zealand's Resource Management Act. These visioning frameworks should elevate local culture, Indigenous knowledge, and youth perspectives.*

Integrate Climate Adaptation in Schools

K–12 public education must become a key platform for Adaptation. Climate curricula should teach:

- *Place-based resilience strategies*
- *Systems thinking (energy, water, food, health)*
- *Emotional resilience and climate justice*
- *Civic engagement and design-thinking*

Schools should also serve as adaptation models—with green infrastructure, rainwater harvesting, and energy systems students help design and manage.

Fund Community-Based Pilot Projects

- *Governments, philanthropies, and planning agencies should fund small-scale adaptation projects that reflect community blueprints—such as food forests, storytelling events, or green schoolyards. These pilots serve as testbeds for larger transformation and help build public ownership.*

Institutionalizing Blueprint Tools Across Sectors

Municipalities must embed blueprint frameworks across departments (urban planning, parks, education, health), as seen in Copenhagen's citywide Adaptation. Key actions include:

- *Updating zoning and building codes for climate resilience*
- *Embedding Adaptation into master plans and comprehensive plans*
- *Tracking metrics such as stormwater absorption, air quality, and climate literacy*

Legal Foundation

Legal and international frameworks support Blueprint planning:

- *U.S. NEPA and HUD statutes mandate participatory visioning.*
- *UNFCCC Article 7 supports community-based adaptation strategies.*
- *UNDRIP guarantees Indigenous peoples' rights to shape futures that affect their lands.*
- *EU Aarhus Convention and SEA Directive ensure visioning in environmental decision-making.*

Call to Action

We urge legislators, school boards, and municipal planners to adopt flourishing blueprints that integrate climate adaptation, education, and community vision into law and practice. The climate crisis demands more than defense—it demands transformation. Flourishing is not a luxury. It is a right and a responsibility.

Planners Toolkit

Equip planners with practical steps to integrate climate adaptation, education, and community visioning into municipal planning and school infrastructure, creating equitable, resilient, and regenerative communities.

1. Start with Visioning

- *Convene community workshops, youth forums, and Indigenous-led circles to co-create a shared vision. Use maps, art, and storytelling to identify local climate risks, assets, and values.*

2. Assess Local Conditions

- *Conduct climate vulnerability assessments tailored to your region (e.g., heat, flood, fire, drought). Including social vulnerability indices to center equity and health impacts.*

3. Design Adaptive Systems

Coordinate plans across sectors:

- *Energy: solar, battery storage, microgrids*
- *Water: green infrastructure, rain gardens, reuse*
- *Food: school gardens, food forests, local supply chains*
- *Housing: heat-resilient retrofits, zoning for defensible space*
- *Mobility: shaded walkways, safe bike routes, transit hubs*
- *Health & Schools: cooling centers, trauma-informed programs*

4. *Partner with Schools*

Use school grounds as pilot sites for Adaptation. Install gardens, bioswales, and solar panels. Integrate climate into lesson plans through place-based, justice-focused curricula.

5. *Institutionalize the Blueprint*

Embed adaptation goals into general plans, zoning codes, bond measures, and educational standards. Coordinate across agencies (planning, education, health, public works).

6. *Fund and Pilot*

Leverage grants, resilience bonds, and public-private partnerships to launch pilot projects. Document outcomes and refine approaches.

7. *Monitor, Celebrate, Iterate*

Track progress through data and community stories. Utilize school exhibitions, local media, and public art to commemorate milestones and maintain momentum.

Resources

1. *ICLEI – Pathways to Resilient Communities (2024)*
2. *Resilient Cities Network – Visioning for Urban Futures (2025)*
3. *Doughnut Economics Action Lab (DEAL)* https://doughnuteconomics.org
4. *Harvard Flourishing Program – Human Flourishing Measure (2023)* https://hfh.fas.harvard.edu
5. *OECD – How's Life? Wellbeing Framework (2024)*
6. *UNDP – Human Development Report (2024): Thriving on a Changing Planet*
7. *The Wellbeing Economy Alliance (WEAll)* https://weall.org
8. *Project Drawdown – Climate Solutions 101 & Community Visioning Guides* https://drawdown.org
9. *Good Ancestor Movement – Intergenerational Visioning for Resilience*
10. *Indigenous Environmental Network – FPIC and Visioning Guides (2025)*
11. *ICLEI – Legal Frameworks for Climate Visioning (2025)*
12. *David W. Orr – Earth in Mind: On Education, Environment, and the Human Prospect (2004)*
13. *Kate Raworth – Doughnut Economics: Seven Ways to Think Like a 21st-Century Economist (2017)*

Chapter Twelve

Climate Change Impacts on the Natural Environment and Adaptation Measures

"We are facing an artificial disaster on a global scale. Our greatest threat in thousands of years. Climate change is real, and we are the last generation that can do something about it."
— Sir David Attenborough, United Nations Climate Summit, 2018

A. The Natural Environment and the Scale of Climate Change

Climate change is reshaping the Earth's natural systems at an unprecedented rate, disrupting ecosystems, biodiversity, hydrological cycles, and land-use patterns. These environmental transformations have cascading effects on human health, food security, and global stability, prompting the need for urgent adaptation planning at local, regional, and international levels. The air and water systems themselves are changing and will determine many aspects of climate adaptation for the current generation. These two global systems interact with each other and with landmasses, although they are considered separately here.

B. Jet Stream Changes and Climate Change: Causes, Impacts, and Consequences

The jet stream—a high-altitude, fast-flowing air current that circles the globe—is a major driver of weather patterns in the Northern and Southern Hemispheres. It acts like a conveyor belt for storm systems, precipitation, and temperature patterns. However, as the planet warms, the jet stream is becoming more erratic, with profound implications for ecosystems, agriculture, and human health.

Jet streams are narrow bands of strong wind in the upper levels of the atmosphere (typically the tropopause, around 5 to 9 miles above the Earth's surface). The most studied is the polar jet stream, which separates cold polar air from warmer air in mid-latitudes and significantly influences weather in North America, Europe, and Asia.

The Arctic is warming nearly four times faster than the global average—a phenomenon called Arctic amplification. This reduces the temperature gradient between the cold polar air and the warmer mid-latitudes, which in turn weakens the strength of the polar jet stream. A weaker jet stream moves more slowly and can develop larger meanders or "wavy" patterns.

Instead of flowing steadily from west to east, the jet stream can become blocked or form persistent ridges and troughs. These distortions lead to stalled weather systems—such as prolonged heatwaves, cold spells, floods, or droughts—that persist over a single area for days or weeks.

Warming oceans, particularly in the North Atlantic and Pacific, impact the jet stream through alterations in atmospheric pressure systems and sea surface temperatures (SSTs). The El Niño–Southern Oscillation (ENSO) and polar vortex disruptions can also interact with jet stream behavior, especially in winter.

C. Impacts of Jet Stream Changes on Natural and Human Systems

Heatwaves and Droughts

Persistent ridges in the jet stream trap warm air over regions for extended periods. Examples include the 2010 Russian heatwave, the 2021 Pacific Northwest heat dome, and the 2019–2023 European summer droughts. These events destroy crops, overheat cities, and intensify wildfires.

When the jet stream stalls, storm systems may linger over the same region, resulting in excessive rainfall. This contributed to devastating floods in Pakistan (2022), Germany and Belgium (2021), and Midwest U.S. River basins.

Paradoxically, a weakened jet stream can allow cold Arctic air to spill southward in winter. For example, the February 2021 Texas freeze was linked to a disrupted polar vortex and a buckled jet stream pattern.

Shifting jet stream patterns affect growing seasons, rainfall reliability, and pollination cycles. Crops dependent on seasonal stability (wheat, maize, rice) suffer yield reductions from unexpected frosts, floods, or droughts.

Unpredictable temperature and precipitation patterns stress species adapted to narrow climatic ranges. Forest ecosystems are vulnerable to drought-induced fires, insect outbreaks, and slow regeneration under altered jet stream dynamics.

Heatwaves increase hospitalization and mortality, especially among the elderly and low-income populations. Floods and extreme storms damage roads, energy systems, and housing, especially in vulnerable urban areas. A study by Francis and Vavrus (2023) concluded that Arctic warming has significantly increased the frequency of extreme weather patterns in the Northern Hemisphere due to jet stream distortions. The World Meteorological Organization (2024) linked unusual summer flooding in Asia and Europe to the stalled jet stream.

D. Changes in Ocean Currents and Their Climate Impacts

Ocean currents play a crucial role in regulating the Earth's climate by redistributing heat, carbon, nutrients, and moisture globally. However, climate change is disrupting these vast "conveyor belts," with consequences that reach from the deep sea to the skies, affecting marine life, weather patterns, food security, and coastal populations.

Ocean currents are large-scale movements of seawater, driven by a combination of wind, Earth's rotation (Coriolis effect), temperature and salinity gradients, and gravitational forces.

The most influential system is the Thermohaline Circulation, also called the Atlantic Meridional Overturning Circulation (AMOC). This global conveyor belt transports warm surface water and cold deep water across the Atlantic, Pacific, and Indian Oceans.

These currents help stabilize global climate by:

- Transporting heat from the tropics to the poles.
- Supporting marine ecosystems through nutrient cycling.
- Regulating rainfall and monsoon systems.
- Absorbing large amounts of atmospheric carbon dioxide.

Arctic and Greenland Ice Melt Weakens the AMOC. Melting ice sheets inject large volumes of cold freshwater into the North Atlantic, lowering salinity and density, which are key drivers of deepwater formation. This slows down the sinking of cold water and weakens the AMOC. The IPCC (2023) projects that the AMOC may decline by 15–50% by 2100 if emissions remain high. Warmer Ocean temperatures reduce the temperature gradient between the equator and poles, which is essential for driving currents. Warming also causes stratification, where lighter, warmer surface water fails to mix with cooler, nutrient-rich water below, disrupting vertical circulation and productivity.

Shifting wind systems (linked to jet stream changes, El Niño, and polar vortex disruptions) alter the direction and strength of surface currents like the Gulf Stream, Kuroshio Current, and Agulhas Current. These changes alter coastal upwelling zones, which are crucial for fisheries and marine biodiversity.

E. Impacts of Changing Ocean Currents

Weakened or altered ocean currents disrupt climate systems across continents: A slowed AMOC is associated with hotter summers and colder winters in Europe, stronger Atlantic hurricanes, and drought in the Sahel and Amazon. The Indian monsoon may become less predictable, posing a threat to water security for over a billion people. El Niño and La Niña events are becoming more frequent and intense, leading to increased global heatwaves, droughts, and floods. Regions like the U.S. East Coast experience accelerated sea level rise when the Gulf Stream weakens because less water is being pulled away from the coast. This can cause sea levels to rise 2–3 times faster than the global average in some locations.

Disruptions in upwelling reduce the flow of nutrients to surface waters, threatening the base of the food web—plankton, which sustain fish, whales, and seabirds. Coral reefs and kelp forests are particularly vulnerable due to ocean acidification, warming, and current shifts that impair larval transport and reproduction.

The ocean has absorbed about 25% of human CO_2 emissions and 90% of excess heat. As circulation slows, this carbon sink capacity is weakening, accelerating global warming. Stagnant waters also risk developing dead zones with low oxygen levels, which can kill fish and invertebrates.

Many of the world's richest fisheries (Peru, West Africa, Northeast Asia) depend on predictable upwelling currents. Changes in ocean dynamics lead to fish migration, population decline, and economic instability for coastal communities. The FAO (2024) warns that 30–40% of commercial fish stocks are at risk due to changing ocean conditions.

IPCC Sixth Assessment Report (2023): Projects a "very likely" weakening of AMOC, with possible tipping points beyond 2°C warming. Nature Climate Change (2024): A study led by Rahmstorf et al. suggests AMOC weakening may already be underway and could collapse this century. NOAA and NASA (2025): Confirmed surface current slowing in the North Atlantic and Pacific using satellite altimetry and ocean buoys.

Programs like the Argo Float Network, GOOS, and NASA's Sentinel satellites monitor temperature, salinity, currents, and sea level to track changes in real-time. Integrated Marine Spatial Planning (MSP) takes into account dynamic ocean conditions, enabling nations to effectively manage fisheries, biodiversity zones, and offshore energy development. Coastal communities are investing in sustainable aquaculture, ecosystem restoration, and diversified livelihoods to reduce dependency on vulnerable marine resources.

F. Impacts on the Natural Environment

Climate change alters temperature and precipitation patterns, shifting species' habitats and threatening extinction for those unable to adapt or migrate. Coral bleaching due to ocean warming, polar habitat loss, and desertification are prominent examples. The Intergovernmental Science-Policy Platform on Biodiversity and Ecosystem Services (IPBES, 2024) estimates over 1 million species are at risk due to climate-driven pressures.

Warmer temperatures, prolonged droughts, and pest infestations (e.g., bark beetles) are degrading forests across North America, Australia, and the Amazon basin. These weakened forests are more prone to wildfires, which further release carbon and reduce nature's capacity to absorb carbon—creating a feedback loop that accelerates warming.

Oceans absorb over 90% of excess heat from global warming. This causes thermal expansion (sea level rise), weakens marine food webs, and contributes to coral reef collapse. Acidification from CO_2 uptake impairs shell formation in mollusks and plankton, disrupting entire aquatic ecosystems. Low oxygen zones (dead zones) are expanding in coastal regions due to warming and nutrient pollution.

Glaciers are retreating worldwide, from the Andes and Himalayas to Greenland and Antarctica. This threatens freshwater availability for over 2 billion people. Meanwhile, altered rainfall patterns lead to more frequent floods in some regions and severe droughts in others, straining wetlands, aquifers, and riparian ecosystems.

Rising temperatures and unsustainable land practices accelerate soil erosion, salinization, and desertification, particularly in arid and semi-arid regions such as the Sahel, Central Asia, and parts of the U.S. Southwest. These changes degrade natural habitats, reduce agricultural productivity, and increase vulnerability to dust storms and landslides.

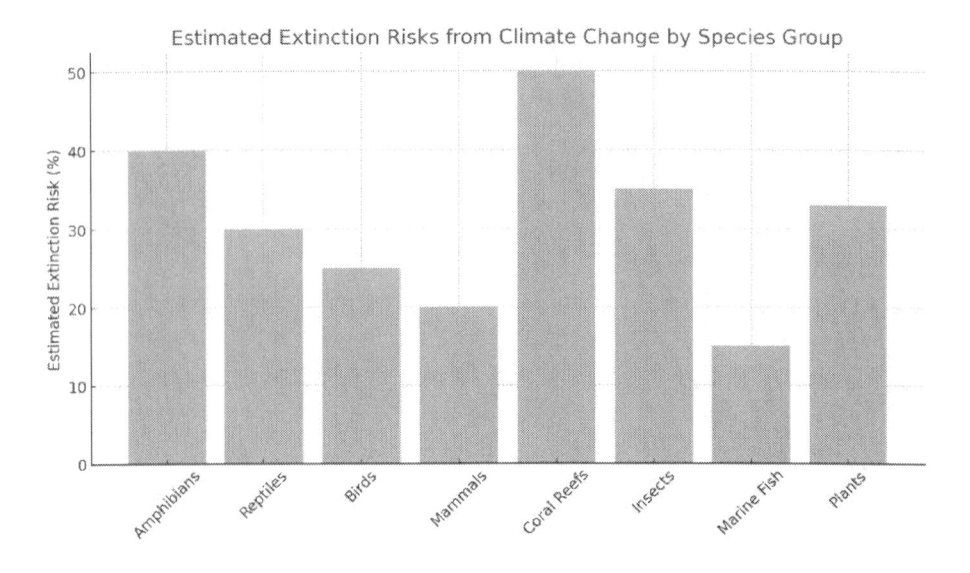

G. Extinctions Due to Climate Change

Climate change is now a leading driver of species extinction, rivaling habitat destruction, overexploitation, and pollution as key factors contributing to the decline of biodiversity. As global temperatures rise, ecosystems are disrupted, species ranges shift, and survival thresholds are exceeded. Many plants and animals are unable to adapt quickly enough, especially those with narrow climate tolerances, small populations, or restricted ranges.

Habitat Loss and Degradation
Rising temperatures, shifting rainfall patterns, and sea level rise are transforming or eliminating critical habitats such as coral reefs, alpine zones, Arctic tundra, wetlands, and cloud forests. For instance, coral bleaching caused by marine heatwaves is wiping out reef ecosystems that support over 25% of marine biodiversity.

Disrupted Life Cycles and Mismatched Timing
Many species rely on seasonal cues, such as temperature and rainfall, for reproduction, migration, or feeding. Climate change causes phenological mismatches, such as birds migrating after insects have peaked or flowers blooming before pollinators arrive. This can collapse food webs.

Temperature Extremes Exceeding Tolerance Limits
Species adapted to narrow temperature bands—such as amphibians, reef-building corals, or high-altitude mammals—can face physiological stress or die-offs when temperatures exceed survival limits. According to the IPCC (2023), every fraction of a degree of warming increases the number of species at risk.

Range Contraction and Isolation
Species are migrating toward the poles or higher altitudes to stay within climate envelopes. But habitat fragmentation, urban barriers, and finite mountain elevations limit migration, leading to range contractions or "mountaintop extinctions."

Ocean Acidification and Deoxygenation
Increased CO_2 lowers ocean pH, impairing calcification in corals, shellfish, and plankton. At the same time, warming leads to the expansion of oxygen minimum zones, suffocating marine life. These combined stressors are pushing entire ecosystems toward collapse.

Current and Projected Extinctions

IPBES Global Assessment (2024): Over 1 million species are projected to face extinction within decades due to the combined effects of human pressures, with climate change rapidly becoming the dominant threat. The IUCN Red List (2025) lists climate change as a primary threat to more than 12,000 endangered species, including iconic animals such as the polar bear, African penguin, and numerous amphibians. Projected Losses: A 2°C rise in global temperature could threaten up to 18% of all terrestrial species, while a 4.5°C rise could imperil up to 50% of species in biodiversity hotspots such as the Amazon, Congo Basin, and Southeast Asia.

H. Case Studies of Climate-Linked Extinction

Bramble Cay Melomys (Melomys rubicola)

- Declared extinct in 2016, this small rodent from a low-lying island in Australia's Torres Strait was the first recorded mammalian extinction directly attributed to climate change—specifically sea level rise and storm surges.

Golden Toad (Incilius perigones)

Once abundant in Costa Rica's cloud forests, the golden toad vanished after a series of arid and warm years in the late 1980s, which altered its breeding habitat and enabled disease spread (chytridiomycosis).

Coral Reef Species

- Entire reef systems are collapsing under the combined pressure of warming, acidification, and bleaching. In Australia's Great Barrier Reef, more than 50% of the coral cover has died since 1995, affecting thousands of species that depend on it.

I. Ripple Effects: Ecosystem Collapse and Human Impacts

Species extinction triggers trophic cascades and ecosystem collapse:

- The loss of pollinators (bees, bats, birds) threatens global food systems.
- Declines in fish populations imperil protein sources for billions of people.
- Disappearing forests and wetlands reduce carbon storage and flood protection.

These extinctions disproportionately affect Indigenous communities, smallholder farmers, and coastal populations, whose livelihoods depend on stable, functioning ecosystems.

Prevention and Adaptation Strategies

Assisted Migration and Genetic Rescue

- Conservationists are relocating species or introducing genetic diversity to help them adapt to changing climates.

Climate-Informed Protected Areas

- Expanding and connecting protected areas based on future climate models allows for range shifts and refugia protection.

Nature-Based Solutions

- Restoring ecosystems, such as mangroves, peatlands, and forests, builds resilience for both biodiversity and people.

Climate Mitigation and Emission Reductions

- The best long-term strategy is reducing greenhouse gas emissions. Every degree of avoided warming reduces the number of species lost.

J. Adaptation Plans and Strategies

Some AI programs use biodiversity and ecosystem services to reduce climate risks. Examples include restoring mangroves to buffer storm surges, preserving wetlands to absorb floods, and reforesting watersheds to stabilize soil and improve water retention. These nature-based solutions offer co-benefits for biodiversity and livelihoods.

Integrating climate forecasts into zoning, conservation easements, and land use planning is essential. Buffer zones, wildlife corridors, and agroecological zoning help species migration and protect ecological integrity. Indigenous knowledge and biocultural landscapes are increasingly recognized in conservation adaptation plans (UNEP, 2024).

Protected areas are being redesigned to account for future climate projections, ensuring species can shift ranges as necessary. Climate refugia—areas buffered from extreme climate impacts—are prioritized for conservation investment, especially in mountainous and coastal transition zones.

Watershed-scale planning, groundwater recharge projects, desalination, and water recycling are key adaptation tools. Natural infrastructure, such as green roofs, bioswales, and permeable surfaces in urban areas, mitigates flood risks and improves water infiltration.

Coastal nations are adopting marine protected areas and marine spatial plans to safeguard biodiversity and sustain fisheries under climate stress. "Blue carbon" ecosystems—mangroves, salt marshes, seagrasses—are restored to enhance carbon sequestration and coastal defense.

When species cannot migrate fast enough or face habitat fragmentation, conservationists may intervene through assisted migration—translocating species to suitable climates. This is a controversial but increasingly considered approach for high-risk taxa, such as alpine plants and amphibians.

Satellite imaging, sensor networks, and artificial intelligence are revolutionizing the way environmental changes are tracked and managed. AI models help identify critical habitats, simulate ecosystem shifts, and support real-time decision-making for land and resource management (Cruz & Kumar, 2024; Google Earth Engine, 2025).

Artificial Intelligence (AI) is becoming a powerful ally in protecting the natural environment from the impacts of climate change. It helps monitor ecosystems, predict environmental threats, optimize conservation strategies, and enhance adaptive management. Below is a narrative overview of how AI is being applied to safeguard nature in the face of climate stress.

K. AI Applications to Protect the Natural Environment from Climate Change

Artificial Intelligence is revolutionizing how scientists, conservationists, and planners respond to ecological degradation caused by climate change. Through advanced data processing, machine learning, and predictive analytics, AI enables rapid, scalable, and often cost-effective interventions to protect the most vulnerable systems in nature.

Ecosystem Monitoring and Early Warning Systems

AI-driven remote sensing tools, such as satellite imagery and drone footage, are used to detect deforestation, glacier retreat, coral bleaching, wetland degradation, and illegal land use in near real-time. Machine learning algorithms classify land cover types and detect anomalies such as heat stress in forests or changes in coastal vegetation.

Engine and Microsoft's AI for Earth help track environmental changes across vast landscapes.

Predictive Habitat and Species Modeling

AI models predict how climate change will affect biodiversity by simulating species range shifts, migration corridors, and extinction risks under various climate scenarios. These tools guide conservation efforts toward high-priority zones such as climate refugia, where ecosystems are buffered from extreme changes. The Nature Conservancy's Resilient Lands Mapping Tool uses AI to identify these zones across North America.

Climate-Resilient Conservation Planning

AI supports conservation planners in designing protected areas and corridors that account for future climate conditions. Algorithms optimize land acquisition or restoration sites for biodiversity preservation, carbon storage, and water security. In Africa, AI-based spatial prioritization tools are helping governments expand nature reserves while enhancing resilience to droughts and floods.

Forest and Carbon Monitoring

Machine learning tools detect patterns of forest degradation, evaluate wildfire risk, and quantify the potential for carbon sequestration. In Brazil, MapBiomas and the National Institute for Space Research (INPE) utilize AI to combat illegal logging and monitor forest loss across the Amazon. AI systems also support REDD+ programs that pay for carbon conservation.

Blue Carbon and Ocean Protection

AI applications are mapping and monitoring "blue carbon" ecosystems—mangroves, salt marshes, and seagrasses—that store significant amounts of CO_2. AI detects ocean warming, acidification trends, and deoxygenation events that threaten marine biodiversity. NOAA's AI models, for example, predict coral bleaching events weeks in advance, enabling rapid responses from marine park managers.

Invasive Species and Pest Management

Climate change is shifting the ranges of invasive species and disease-carrying pests. AI helps track their movement, forecast outbreaks, and guide response efforts. In agriculture and forestry, AI systems detect infestations (like bark beetles in conifer forests) and recommend intervention strategies to prevent ecosystem collapse.

Environmental Law and Governance Tools

Natural language processing (NLP) is being utilized to analyze environmental impact statements, compliance documents, and climate policies, identifying gaps or violations in conservation laws. AI can also monitor social media and satellite data to detect illegal activities such as poaching or mining in protected areas.

L. Conclusion

The impacts of climate change on the natural environment are significant and severe. They are essential to the Climate Adaptation generation because they are so noticeable in their extremes.

Advocacy Brief

Climate change is accelerating the global extinction crisis. Rising temperatures, shifting rainfall patterns, ocean acidification, and habitat disruption are driving thousands of species toward extinction at rates not seen since the last mass extinction event. The loss of biodiversity poses a significant threat to food security, ecosystem stability, and human health.

Key Facts:

- *Over 1 million species are at risk of extinction, with many facing extinction within decades (IPBES, 2024).*
- *Climate change is now listed as a significant threat to over 12,000 species on the IUCN Red List (2025).*
- *The Bramble Cay Melomys was the first mammal to go extinct directly due to sea level rise.*
- *Coral reef systems, vital to marine biodiversity and coastal economies, have lost over 50% of their cover since the 1990s.*
- *A global temperature rise of 2°C could result in the loss of up to 18% of terrestrial species, while an increase of 4.5°C could push that to 50% in key ecosystems.*

Call to Action:

1. *Integrate biodiversity protection into climate policy at all levels of government.*
2. *Support climate-adaptive conservation programs, including habitat corridors and species relocation initiatives.*
3. *Strengthen climate mitigation efforts to limit warming to 1.5°C, which could significantly reduce extinction risks.*
4. *Support Indigenous stewardship, which preserves approximately 80% of the world's biodiversity.*

5. Adopt nature-based solutions that protect both ecosystems and communities.:

Extinction is irreversible. Immediate action is necessary to reduce emissions, protect habitats, and plan for climate-adapted conservation. Protecting nature is not just about saving wildlife—it's about safeguarding the future of humanity.

Planners Toolkit

Addressing Extinction from Climate Change

Urban and regional planners play a vital role in safeguarding biodiversity under climate stress. As ecosystems face rapid change, planners can integrate adaptive strategies into land use, development, and environmental review processes to prevent extinction and protect ecosystem services.

1. Climate-Resilient Zoning and Land Use

- *Prioritize conservation of climate refugia (areas buffered from extreme climate change).*
- *Designate green corridors to enable species migration.*
- *Use overlay zones to limit development in biodiversity hotspots.*

2. Environmental Impact Assessments (EIA)

- *Require climate vulnerability assessments for proposed developments.*
- *Include species-at-risk screening using the IUCN Red List and regional biodiversity data.*
- *Mandate cumulative impact analysis to prevent piecemeal habitat loss.*

3. Nature-Based Infrastructure

- *Restore wetlands, riparian buffers, and green roofs to support habitat and climate resilience.*
- *Incorporate native species and pollinator pathways into urban design.*
- *Use green belts and agroforestry buffers to connect fragmented habitats.*

4. Conservation and Land Acquisition Tools

- *Partner with land trusts and Indigenous organizations to conserve key areas of land.*
- *Leverage conservation easements, transfer of development rights (TDR), and mitigation banking.*
- *Use AI-powered planning tools to predict ecosystem shifts (e.g., NatureServe, MapBiomas).*

Toolkit Resources:

- *IUCN Red List Spatial Data:* <u>Toolkit Resources:</u>
- *NatureServe Climate Vulnerability Explorer:* <u>Toolkit Resources:</u>

- *Esri Green Infrastructure Planner: https://www.esri.com*
- *UNEP Urban Nature-Based Solutions Framework: https://www.unep.org*

Resources

1. Cruz, M. J., & Kumar, S. (2024). Artificial intelligence in climate adaptation and environmental resilience. *Climate Intelligence Review, 11(2), 45–61.*
2. Intergovernmental Panel on Climate Change (IPCC). (2023). *Sixth Assessment Report: Impacts, Adaptation, and Vulnerability.* https://www.ipcc.ch/report/sixth-assessment-report-working-group-ii/
3. Intergovernmental Science-Policy Platform on Biodiversity and Ecosystem Services (IPBES). (2024). *Global Biodiversity Outlook: Climate Chapter.* https://www.ipbes.net
4. UNEP. (2024). *Ecosystem-based adaptation: Guiding principles and framework for action.* United Nations Environment Programme. https://www.unep.org
5. Google Earth Engine. (2025). *Ecosystem intelligence through satellite-aided adaptive planning.* https://earthengine.google.com Cruz, M. J., & Kumar, S. (2024). Artificial intelligence in climate adaptation and environmental resilience. *Climate Intelligence Review, 11(2), 45–61.*
6. Google Research. (2025). *AI for coral reef monitoring and blue carbon detection.* https://ai.google/research
7. Microsoft. (2024). *AI for Earth: Empowering environmental sustainability with artificial intelligence.* https://www.microsoft.com/en-us/ai/ai-for-earth
8. NatureServe. (2025). *Predictive biodiversity risk modeling using AI.* https://www.natureserve.org
9. United Nations Environment Programme (UNEP). (2024). *Digital solutions for nature-based climate adaptation.* https://www.unep.org · Rahmstorf, S., Box, J. E., & Smeed, D. A. (2024). Observed weakening of the Atlantic Meridional Overturning Circulation. *Nature Climate Change, 14(2), 103–111.*
10. Food and Agriculture Organization (FAO). (2024). *The State of World Fisheries and Aquaculture.* https://www.fao.org
11. National Oceanic and Atmospheric Administration (NOAA). (2025). *Ocean Current Monitoring and Climate Change.* https://www.noaa.gov
12. NASA Earth Observatory. (2025). *Satellite Data Confirms Ocean Circulation Trends.* https://earthobservatory.nasa.gov IUCN. (2025). *The IUCN Red List of Threatened Species.* https://www.iucnredlist.org
13. Pacifici, M., Visconti, P., & Rondinini, C. (2023). The risk of species extinction due to climate change is accelerating. *Global Ecology and Biogeography, 32(1), 12–25.*
14. Watson, J. E. M., et al. (2024). Extinction risk and climate vulnerability of biodiversity hotspots. *Nature Sustainability, 7(2), 110–118.*

Chapter Thirteen

The Climate Adaptation Revolution and Evolution

"*Ecology is freedom. Without care for land and water, there is no revolution.*"
— *Rojava Environmental Committee (2025)*

A. Is it a Climate Adaption Revolution or Evolution?

The Climate Adaptation Revolution is a transformative shift in how societies confront the intensifying effects of climate change—not merely through reactive measures but through proactive and deeply place-based approaches. Unlike climate impacts mitigation, which focuses on reducing emissions to prevent future climate impacts, Adaptation acknowledges that climate disruptions are already unfolding. Rising sea levels, intensified storms, extended droughts, wildfires, biodiversity and ecosystem collapse, possible earthquakes, and robust public health threats are increasing. Climate Adaptation is cultural, ethical, political, and rooted in collective imagination and action. And the hope for the types of climate adaptation we all need comes from the Climate generation described in the first chapter.

Historically, climate response strategies emphasized risk management through infrastructure protection and economic loss prevention. Today's adaptation revolution redefines Adaptation by inclusion, and the capacity of communities to "bounce forward," not merely "bounce back." As noted in the IPCC Adaptation Report Summary (2023), Adaptation must be reimagined as the ability to transform in ways that enhance fairness, safety, and dignity across generations. In other words, adaptation should be us to aspire to flourish, as described in the earlier chapters, not just return to the status quo.

A defining feature of this revolution or evolution is the rise of localized, place-based strategies that draw on the unique ecological, cultural, and historical characteristics of specific regions. These strategies incorporate traditional ecological knowledge, community memory, and participatory planning to design solutions that fit the landscape and the people who inhabit it. Current practices of land use zoning are not place based.

Governance is also undergoing a profound transformation. No longer confined to top-down emergency response, adaptation planning is increasingly participatory and democratic. Youth, elders, frontline communities, Indigenous peoples, scientists, and local governments are collaborating to co-create visions for resilient futures. The rise on global communication followed the rise in global communication technology.

The United Nations Office for Disaster Risk Reduction (UNDRR) emphasized in its 2024 Global Risk Governance Report that this shift represents a democratic revolution—one that requires listening deeply to those who have been historically marginalized or silenced.

Another hallmark of this change is the integration of nature-based and culturally grounded solutions. These include the restoration of wetlands, mangroves, forests, and watersheds alongside the revitalization of ancestral practices of land stewardship, collective care, and reciprocity. The Indigenous Climate Network's statement at COP29 (2024) captured this ethos powerfully:

> "*To adapt to climate change, we must repair our relationship with nature and each other.*"

Technology also plays a role in the adaptation revolution. Artificial intelligence, satellite imaging, predictive analytics, and early warning systems are transforming the way we assess risk, manage natural resources, and plan for future conditions. As noted by Cruz and Kumar in the Climate Intelligence Review (2024), "AI won't save us from climate chaos—but it can help us organize, visualize, and mobilize."

This revolution or evolution is not theoretical; it is already unfolding. In Bangladesh, communities have created floating farms, schools, and clinics using bamboo and water hyacinths to adapt to chronic flooding. These innovations blend traditional knowledge with modern needs. In California, Indigenous groups, such as the Karuk and Yurok, are leading efforts to reintroduce traditional cultural fire practices that reduce wildfire risks and restore ecological balance, partnering with state agencies. In the Netherlands, the "Room for the River" program has radically redesigned urban spaces to accommodate floodwaters through parks, green zones, and ecological corridors. In South Africa, resilience hubs in Cape Town and Durban offer solar power, clean water storage, emergency response services, and climate education—particularly in historically underserved neighborhoods.

Underlying these real-world adaptations is a broader demand for systemic change. Policy frameworks, ranging from zoning codes to environmental law, must be revised to reflect the realities of a warming world, as outlined in the other chapters.

Climate finance must shift away from carbon-centric metrics toward investments in frontline community leadership, nature-based infrastructure, and human-centered design. Education must incorporate climate literacy and collective visioning as core components of curriculum—from primary school to adult learning. My book, "Climate Change in the Classroom: Celebrating Optimism" and chapter nine here go into exactly how to bring climate education into the classroom.

Healthcare systems must be redesigned to address emerging climate health challenges, including heat stress, vector-borne diseases, and climate-related mental health impacts

In many parts of the world, governments have been slow, fragmented, or politically constrained in their responses to climate disruption, as discussed in the next chapter. National adaptation plans, while important, are frequently reactive, top-down, or inadequately resourced. As a result, the communities are experiencing climate change impacts —drought, flood, rising seas, wildfire, storm intensification, or ecosystem collapse— are at the forefront of acclimate adaptation development.

These localized, nongovernmental efforts go in many directions. In the Amazon Basin, Indigenous federations such as COICA are mapping climate-vulnerable territories, restoring forests, and asserting ancestral land rights as a core strategy of Adaptation. In urban United States, frontline coalitions like the West Harlem Environmental Action Commission or Climate Justice Alliance are developing community-owned solar grids, resilience hubs, and food sovereignty systems in neighborhoods that have been long excluded from federal investment. In the Sahel region of Africa, farmers are reviving traditional water-harvesting techniques and establishing cooperative seed banks without waiting for state-directed climate change programs. These are not merely acts of Adaptation; they are declarations of sovereignty, agency, and cultural survival.

The shift also encompasses transboundary networks and new alliances that transcend government boundaries. City-level coalitions, such as C40 Cities and ICLEI—Local Governments for Sustainability, enable municipalities to coordinate adaptation practices independently of national politics. Faith communities, youth climate movements like Fridays for Future, and regional mutual aid groups have become critical infrastructure in the face of climate disasters, often outpacing the reach or speed of necessary emergency responders' agencies.

The climate adaptation revolution, then, is not government-led—it is government-inclusive and considered by some to be more an evolution. But its center of gravity has shifted. It lives in the hands of people who plant mangroves, recharge aquifers, organize care networks, rewrite zoning codes, and teach future generations how to flourish in a changed world. It is in these acts of localized autonomy, mutual learning, and cross-border solidarity that climate adaptation truly transcends governments and becomes a planetary means of survival and hope.

Next, we examine two ways of thinking about how to make decisions, from capitalism to the Marxism. These are generally considered opposite forms of governance. Let Nature be the judge.

B. Climate Adaptation and Capitalism: Tensions and Changes

As the world becomes increasingly anxious with the mounting impacts of climate change, the relationship between climate adaptation and capitalism has become increasingly central and increasingly contested. On the one hand, market-driven systems are mobilizing massive financial resources, technological innovations, and private-sector involvement in Adaptation. On the other, the structural inequalities, ecological degradation, and extractive logic of capitalism are among the root causes of the climate crisis itself. This paradox lies at the heart of the debate: Can capitalism adapt to climate change, or must climate adaptation evolve beyond capitalism?

The Capitalist Imperative and Market-Based Adaptation

Contemporary climate adaptation has not escaped the reach of global capital. Insurance firms, hedge funds, agribusiness, and technology giants are positioning Adaptation as the next frontier of investment. Adaptation finance, considered a humanitarian issue, is increasingly commodified through climate bonds, green infrastructure portfolios, and resilience-as-a-service models. According to the World Bank (2024), private adaptation financing now exceeds public adaptation funding in over 40% of developing countries.

Real estate developers are building luxury "resilient enclaves" with private seawalls, floodproof towers, and microgrid energy systems. At the same time, marginalized neighborhoods lack access to basic disaster recovery funds.

Tech firms market AI-driven adaptation platforms but often charge licensing fees beyond the reach of most municipalities. Agribusinesses promote genetically engineered drought-resistant crops while displacing smallholder farmers and degrading seed sovereignty.

> "Capitalism does not just adapt to climate disruption, it profits from it. The question is, who is left behind?"
> — Naomi Klein, Climate Author and Scholar (2025)

This dynamic creates what many call "climate apartheid"—where the wealthy buy their way to safety while the poor are left exposed. As UN Special Rapporteur Philip Alston warned, "Climate change threatens to undo the last fifty years of development, global health, and poverty reduction... yet business-as-usual adaptation reinforces existing inequality." (UN Human Rights Council, 2024)

Adaptation Beyond Capitalism

Increasing numbers of argue that meaningful Adaptation cannot occur within the confines of a system built on perpetual extraction and accumulation. They assert that capitalism, by design, externalizes environmental harm and prioritizes short-term profit over long-term planetary survival. Under such a model, Adaptation becomes selective, transactional, and technocratic—rather than collective, ecological, and just.

Grassroots movements in the Global South, Indigenous nations, and climate justice networks are advancing alternative adaptation pathways rooted in commons governance, bioregional planning, agroecology, and relational worldviews. These models challenge the commodification of water, land, knowledge, and care. They advocate for de-growth, solidarity economies, and rights-based approaches to Adaptation that prioritize dignity over profit.

Many of these approaches are not new—they revive pre-capitalist and non-Western traditions of reciprocal stewardship, cooperative labor, and ecological belonging. Adaptation practices that have persisted for centuries often stand in sharp contrast to market-driven models.

The Future: Reconciliation or Rejection?

The path forward is deeply contested. There is a massive amount of sheer power and momentum in any status quo that is entrenched with power. Some argue for a softer approach using state power to regulate markets, redirect capital, and enforce equity through adaptation investments, green jobs, and climate public works.

It is increasingly recognized that if climate adaptation becomes a tool for entrenching inequality, privatizing nature, and displacing vulnerable populations, then it will fail not only ethically but practically. A world of fortified financial districts surrounded by flooded favelas, drought-stricken villages, and burning forests is not an adaptation—it is a collapse in disguise.

The Climate Adaptation generation asks more profound questions: Who owns safety? Who decides the future? What kind of economy sustains life on a damaged planet?

C. Climate Adaptation and Marxism: Reclaiming the Economy in an Age of Crisis

Marxism views climate adaptation—not simply as a set of policies for survival but as a struggle over the ownership, control, and direction of life on a damaged planet. While mainstream adaptation frameworks often focus on technical solutions and market-based responses, a Marxist analysis asks questions: Who benefits from Adaptation? What structural forces produced climate vulnerability in the first place?

Traditional Marxism understands the climate crisis, including the need for Adaptation—because of the capitalist mode of production, which organizes nature and labor around the pursuit of profit, not human or ecological well-being. Under capitalism, Adaptation is increasingly commodified—privatized seawalls, climate-resilient luxury developments, proprietary AI-driven risk platforms, and patented drought-resistant seeds—while the global working class, peasant communities, those on the wrong side of the digital divide, and Indigenous nations face the brunt of rising temperatures, wildfires, storm intensification, flooding, cumulative toxic exposures, and displacement.

In a Marxist view, Adaptation is not just environmental planning; it is a class struggle over land, labor, and life itself.

From coastal fishing communities fighting against land grabs by tourism developers to urban renters resisting green gentrification and displacement, adaptation decisions determine whose lives are protected, whose losses are ignored, and whose futures are prioritized.

This is especially visible in climate-vulnerable regions of the Global South, where adaptation funds are mediated through development banks and aid institutions that often impose market reforms, debt regimes, or conditional privatization of natural resources. For Marxist scholars, this is a continuation of capitalism through Adaptation.

Historical Materialism and Ecological Time

Marxism also emphasizes historical materialism—the understanding that current climate vulnerabilities are not natural but the product of social structures and historical processes: colonization, enclosure, forced labor, slavery, extractive infrastructure, and little or no environmental planning. Adaptation must address how centuries of exploitation have created structural climate risk in certain places and populations.

Toward an Eco-Socialist Adaptation

The Marxist vision of Adaptation is ultimately an eco-socialist one. It sees the climate crisis as a turning point. The contradiction between capital and ecology can no longer be managed. Eco-socialist Adaptation rejects false solutions, such as carbon markets or geoengineering, and instead champions a just transition to democratic control of adaptation planning, energy, agriculture, housing, and infrastructure.

Marxism does not view Adaptation as a policy checklist—it sees on a more pervasive scale. In their view climate adaptation societies must "adapt" through profound transformation, reclaiming land, labor, and ecological time from capital. In this view, if the profound transformation does not occur then Adaptation will become yet another product leaving behind a trail of of dispossession and inequality.

These examples show that climate adaptation informed by Marxist principles is not hypothetical—it is already happening in many locations. Whether in Indigenous territories, socialist cooperatives, revolutionary zones, or peasant uprisings, Adaptation is being reclaimed as a struggle against capital, colonialism, and enclosure. It is being built not for profit but for dignity, memory, and ecological survival.

What unites these many eco socialist movements is a shared rejection of commodified resilience and a commitment to systemic change—where Adaptation is no longer about protecting the status quo but about dismantling it.

D. Generational and Class Differences in AI: An Impediment to Climate Adaptation?

The use of artificial intelligence (AI) varies significantly across age groups, influenced by generational differences in technology adoption, access, education, and trust. These differences impact on how individuals interact with AI tools in their daily lives, at work, in healthcare, and civic engagement. These differences also impact how quickly AI will be utilized to transition into an effective climate adaptation mode. Tying these differences to the power of each group is a measure of an obstacle to climate adaptation. Adding the digital divide as an overlay reveals another layer of challenges to climate change adaptation. Here's a breakdown of key differences:

1. Youth and Digital Natives (Ages 10–25)

This generation has never known a world without the internet—and now they're growing up immersed in AI. From generative art to TikTok algorithms, AI is embedded in their education, entertainment, and social interaction. This is a big part of the Climate Generation. And they are part of the surge of technological development in Climate Change Adaptation.

Youth use AI tools not only for personal convenience but also for creative expression and activism. AI platforms like ChatGPT and Midjourney are widely used for schoolwork, coding, and digital content creation. They are also more comfortable engaging with emerging tech and participatory platforms.

> *"Today's youth are not just consumers of AI—they are shaping its evolution. Their intuitive fluency is rewriting the boundaries of education, communication, and civic participation."*
> — Sandra Wachter, Oxford Internet Institute (2024)

However, concerns around AI-enabled misinformation, privacy breaches, and dependency are growing. The World Economic Forum (2025) notes that digital-native generations are

2. Working-age adults (Ages 26–64)

This broad group straddles early adopters and reluctant users. They utilize AI in workplace tools such as Microsoft Copilot, Zoom AI, and Salesforce Einstein. In healthcare, law, logistics, finance, and education, AI is automating repetitive tasks, personalizing services, and forecasting trends.

Yet this generation also shows significant anxiety over automation and job security. According to a Pew Research Center survey (2025), 62% of working-age adults worry that AI will eliminate more jobs than it creates. Education and retraining are crucial for bridging this gap.

The use of AI in this age group also reflects occupational disparities: tech workers and professionals are far more likely to use AI effectively than those in service, labor, or caregiving roles. Some of the group are on the side of the digital divide without access.

3. Older Adults and Seniors (65+)

Older adults often adopt AI for health monitoring, companionship, and accessibility. Voice interfaces (e.g., Siri, Alexa), AI-powered wearable tech (e.g., Fitbit, Apple Watch), and home assistants offer mobility, medication reminders, and emergency support.

However, usage is limited by the digital divide. Many seniors' express skepticism or discomfort with AI, citing concerns about data privacy, a lack of digital literacy, and worries that human connection will be replaced. Many simply do not cannot trust it. The speed and accuracy of AI response can be overwhelming to people in this cohort. Add to those the higher rates of aged, induced disability in this group.

As we design AI for the future, we must include those who remember the past. Seniors hold generational memory and moral clarity.

> "*AI must be co-designed across generations. Otherwise, we risk building a future that works only for some and leaves others behind.*"
> — *Future of Life Institute AI Ethics Brief (2025)*

E. Digital Divide and Climate Adaptation: Unequal Access in an Era of Crisis

The advance of technology has more evolutionary not revolutionary. Even though its uses are surging, there are those left behind, for now. The digital divide—the gap between those with reliable access to digital technologies and those without—has become a defining fault line in the era of climate change. As artificial intelligence, satellite systems, predictive analytics, and climate information platforms increasingly shape how societies respond to climate risks, communities without access to digital infrastructure, data literacy, and online tools are being left behind. Climate adaptation is no longer just about physical infrastructure and ecological resilience; it now also depends on the ability to connect, access, and apply digital knowledge. This means that the digital divide is becoming a core determinant of who survives and who thrives as climate impacts accelerate.

The relationship between climate adaptation and digital inequality is especially stark in the Global South, Indigenous territories, informal urban settlements, rural areas, and among elderly or low-income populations worldwide. In these areas, many communities lack access to broadband internet, functioning mobile networks, climate data portals in local languages, or the necessary technological skills to navigate weather modeling platforms, risk maps, or early warning systems. While wealthier countries invest in AI-enhanced flood forecasts, wildfire simulations, and remote-sensing technologies, digitally excluded populations must rely on limited analog tools or informal community knowledge to prepare for and respond to disasters.

This asymmetry creates climate information divide. For example, smallholder farmers in Sub-Saharan Africa, Latin America, or South Asia often lack access to or the means to afford precision agriculture tools or climate-adaptive seeds promoted through app-based platforms. Fisherfolk who rely on shifting ocean patterns are rarely included in ocean-atmosphere data networks. Urban residents in slums may never receive digital heatwave alerts because they lack smartphones or are not digitally registered. The very digital systems that could support inclusive adaptation planning often reinforce existing social, geographic, and racial inequalities.

As climate funding increasingly favors "data-driven" Adaptation or "smart city" resilience projects, those without digital capacity are deemed less eligible for investment.

Donors and governments often require data dashboards, GIS maps, and digital reporting for adaptation funding, inadvertently sidelining Indigenous, oral, or locally embedded knowledge systems that may not be digitized. In this way, the digital divide shapes not just how climate information flows but who is heard, who is funded, and whose knowledge is considered valid.

During disasters, the stakes become even higher. Real-time crisis management increasingly relies on AI platforms that analyze social media, sensor networks, and emergency call data to prioritize evacuation, food delivery, and rescue operations. However, communities without digital presence or connectivity are rendered invisible to these systems. Knowledge in this sense is truly power. In climate-vulnerable regions such as the Caribbean, many Pacific Islands, Sub Saharan Africa, or rural Central Asia, being offline can mean being uncounted, unprotected, and unprepared.

Despite these challenges, community-driven solutions are emerging. Organizations like Praxis India are building inclusive data platforms using participatory mapping. Indigenous Climate Action in Canada promotes digital sovereignty for First Nations through community-controlled monitoring systems. African youth-led networks, such as the Climate Smart Agriculture Youth Network, are creating open-source climate tools tailored to rural contexts. These initiatives demonstrate that bridging the digital divide is about power, governance, and self-determination in a climate-altered world.

As climate change intensifies, Adaptation will increasingly depend on digital infrastructure. But unless this infrastructure is democratized, built with and for marginalized communities, the digital divide will become a new frontier of climate injustice. Bridging this divide requires investment in public internet, multilingual platforms, digital education, inclusive design, and the recognition of non-digital forms of knowledge as equally valid. It also means asking not just who uses AI or data—but who owns it, who controls it, and who serves it.

The technologies meant to help us adapt may only deepen the very divides we need to overcome.

F. Conclusion

A cross the spectrum of political ideologies from capitalism to Marxism is an undeniable engagement with the changes in the climate.

There is a strong desire everywhere in the Climate Adaptation generation to move past governmental barriers to action plans for how we can really adapt to the known and yet unknown impacts of climate changes.

Advocacy Brief

Issue:

As climate impacts intensify—through extreme heat, drought, flooding, and migration, communities must access timely data, alerts, planning tools, and decision-making platforms to adapt. Yet the global digital divide threatens to turn climate adaptation into a new domain of inequality. Marginalized populations—especially in the Global South, rural areas, Indigenous lands, and low-income urban zones—often lack the digital infrastructure, tools, and skills needed to participate in data-driven adaptation strategies.

Why It Matters:

Adaptation is increasingly digital. Early warning systems, AI-enabled risk assessments, satellite climate models, and intelligent infrastructure guide emergency responses and resource allocation. However, without equitable digital access, communities remain invisible to funding agencies, planning systems, and crisis networks. The result is "climate information poverty"—the phenomenon where people most affected by climate change are least connected to adaptation tools.

Policy Recommendations:

1. *Fund digital public infrastructure in climate-vulnerable regions, including broadband, off-grid connectivity, and mobile weather services.*
2. *Support community-controlled data systems that integrate local knowledge and promote digital sovereignty.*
3. *Ensure the multilingual and culturally responsive design of digital climate tools for rural, Indigenous, and underserved users.*
4. *Make adaptation funding contingent on digital inclusion and participation of low-connectivity communities.*
5. *Train local youth and frontline workers in climate tech tools and AI ethics to build community-led digital capacity.*

Action

Digital access is now a pillar of climate justice. Governments, donors, and climate institutions must invest in closing the digital divide as a prerequisite for equitable Adaptation. No community should be left behind because it cannot log in.

Planners Toolkit

Purpose:

This toolkit supports local planners, municipal officials, and adaptation teams in ensuring that climate resilience efforts are digitally inclusive and equitable—especially in underserved, rural, and Indigenous communities.

1. Assess Local Digital Capacity

- Conduct a digital access audit, including internet availability, mobile coverage, and device access.
- Identify low-connectivity zones vulnerable to climate risks.
- Map who is excluded—by income, age, gender, language, or geography.

2. Co-Design with Communities

- Use participatory tools to engage residents in identifying climate and digital needs.
- Integrate oral knowledge, analog warning systems, and non-digital practices.
- Prioritize youth and digital facilitators, as well as local tech leaders.

3. Build Inclusive Digital Infrastructure

- Advocate for public Wi-Fi, solar-powered hubs, and climate-resilient communication systems.
- Partner with libraries, schools, and health centers as digital adaptation points.
- Enable offline access to weather alerts, adaptation plans, and maps.

4. Promote Digital Climate Literacy

- Train community members in the use of weather apps, climate dashboards, and AI tools.
- Create visual, low-literacy materials in local languages.
- Collaborate with NGOs, universities, and Indigenous networks for technical support.

5. Embed Equity in Planning Tools

- Choose platforms that allow localized data entry and ownership.
- Require open-source and transparent data systems in adaptation contracts.
- Track digital inclusion as a performance metric in resilience planning.

Quick Checklist for Digital Equity in Adaptation Plans

☐ *Does your climate plan reach unconnected populations?*
☐ *Are adaptation tools multilingual and culturally relevant?*

☐ *Are community members co-designing digital strategies?*
☐ *Is local digital literacy being actively supported?*
☐ *Are non-digital solutions still part of your toolkit?*

Remember:

Digital exclusion = climate risk. Equitable Adaptation requires that all communities can see the future coming—and shape how they respond.

Resources

1. Indigenous Climate Action. (2024). Land, water, and justice: Indigenous-led climate adaptation practices. *https://www.indigenousclimateaction.com*
2. Karriem, A., & Satgar, V. (2025). Climate adaptation and democratic eco-socialism: Lessons from the South African Climate Justice Charter. Global South Politics, 4(2), 111–129.
3. Malm, A. (2023). Fossil Capital: The rise of steam power and the roots of global warming. Verso.
4. Movimento dos Trabalhadores Rurais Sem Terra (MST). (2024). Agroecology and land reform as climate adaptation. MST Education Department. *https://mstbrazil.org*
5. Rojava Environmental Committee. (2025). Ecology, autonomy, and resilience: Adaptation in the Democratic Federation of Northeast Syria. Rojava Information Center.
6. Satgar, V. (Ed.). (2024). The climate crisis: South African and global eco-socialist alternatives (2nd ed.). Wits University Press.
7. UNDRR. (2024). Global risk governance: A new framework for decentralized climate resilience. United Nations Office for Disaster Risk Reduction. *https://www.undrr.org*
8. Zapatista Army of National Liberation (EZLN). (2023). Resistance and autonomy in the face of storms: Communiqués from Chiapas. Chiapas Independent Press.

Chapter Fourteen
CONCLUSION

Climate Adaptation: Summary and a Blueprint for Flourishing

> " We live in a borderless crisis with bordered solutions. Until we break that contradiction, we cannot adapt with justice."
> — UN Special Rapporteur on Human Rights and Climate Displacement (2024)

A. Our Journey Together

This book is designed to empower you, particularly the Climate Adaptation generation. The rapid rise of global communication and artificial intelligence is giving us the tools to move forward rapidly. In Chapter One, we examined the demographics of the climate adaptation generation and the older generation who evidenced resilience in their lives when faced with difficult situations. I have not emphasized the term 'climate resilience' because it implies an ability to withstand climate impacts by returning to the status quo rather than embracing a forward-thinking new way of climate adaptation that involves envisioning a direction towards flourishing. The political and economic power structures of each group were identified, and in doing so, we recognize the need to empower the climate adaptation generation to establish a new climate adaptation framework. Chapters two through seven describe the specific natural disasters that can be expected from climate change and how to advocate for and plan for them. I specifically emphasized planners because they often hold the key to opening government direction. In many countries, Planners are the equivalent of public-sector real estate brokers. They typically aim to expand the tax base to boost municipal revenues. They do not traditionally engage in public health, education, environmental protection, or place-based planning. They do not consider visioning or flourishing.

The Climate Adaptation generation will demand Planners proceed with all these to plan for climate adaptation successfully. Chapters Two through Twelve provide them with the necessary tools to do so and equip the Climate Adaptation generation with the information to help Planners move in these directions. Chapter Eight describes what Place-based planning is and its evolving global trend. The raw logic of planning for a place is undeniable when it comes to climate adaptation. How can one plan for Climate Adaptation without considering the specific place? Chapter Nine is a general overview of the role of education. I have detailed this approach in my book, "Climate Change in the Classroom: Celebrating Optimism," where I focus on how to create knowledgeable leaders in climate adaptation. Chapter Ten introduces us to public health issues, which are numerous. Climate anxiety runs rampant and is stultifying forward progress.

Because of the actual public health degradation from climate impacts, this anxiety is not misplaced and is not a sign of mental illness. Public health impacts are real and robust. Chapter eleven demonstrates the relationship between visioning and flourishing. Visioning is a way to reach a place we haven't been before, one that flourishes with climate adaptation. We need the energy and focus of the Climate Adaptation generation to move beyond a return to the status quo and empower them to create a new vision of a better goal – one of flourishing. No book on Climate would be complete without a discussion of the impacts on the Natural environment, which is in Chapter Twelve. They are growing, and many are unknown. Some species may become extinct before we even know they exist. Chapter Thirteen examines the role of political ideology in climate adaptation, from capitalism to Marxism. The Climate Adaptation generation globally is more interested in results, regardless of these ideologies. Climate adaptation is occurring as a means of survival in certain places, with or without government intervention. We can all learn from these efforts. In this chapter, I discuss three significant challenges: the stress between political boundaries and ecosystems and bioregions, the digital divide between the young and the old, the rich and the poor, and the cumulative impacts of climate change

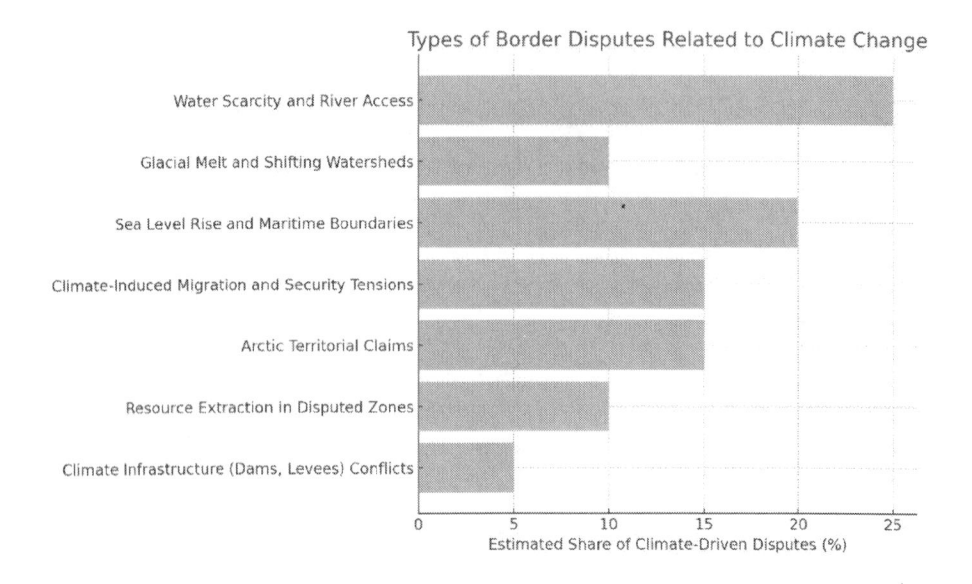

Types of Border Disputes Related to Climate Change

Type	Estimated Share (%)
Water Scarcity and River Access	25
Glacial Melt and Shifting Watersheds	10
Sea Level Rise and Maritime Boundaries	20
Climate-Induced Migration and Security Tensions	15
Arctic Territorial Claims	15
Resource Extraction in Disputed Zones	10
Climate Infrastructure (Dams, Levees) Conflicts	5

Estimated Share of Climate-Driven Disputes (%)

B. Climate Migration: Breaking the Border Paradigm

Marxist theorists argue that Adaptation within a capitalist system is inherently contradictory. The system that created climate chaos cannot provide the transformation needed to survive it. As Andreas Malm writes in Fossil Capital, "The logic of capital is expansion, not Adaptation. It cannot retreat from nature without retreating from itself." This insight prompts us to reimagine Adaptation not as an accommodation to the climate crisis but as a resistance to the economic and political structures that have caused it. One power structure is the political boundary between differing countries, states, and regions.

One of the most profound challenges to national boundaries is climate migration. Rising seas, prolonged drought, and intensifying storms are displacing millions—particularly in island nations, coastal megacities, and fragile agricultural regions. Yet international law remains deeply tied to sovereign borders, offering few protections for climate-displaced or Indigenous persons.

The 1951 Refugee Convention does not recognize Climate as a valid reason for asylum. Countries like the United States, Australia, and many in Europe maintain hardline immigration policies even as their emissions disproportionately contribute to climate vulnerability elsewhere. As rising temperatures and resource scarcity displace millions from their homes, national borders become increasingly hardened. In contrast, transboundary adaptation justice demands a rethinking of climate responsibility, mobility rights, and the moral obligations of high-emitting states.

Over time, many of the environmental impacts of climate change will spread throughout the larger ecosystem. They can migrate through land, air, and water. And now, environmental impacts on water can be detected across the North American continent because we have a better understanding of where all the water exists above and below ground. That is just one example of an expanding knowledge base and one that AI will incorporate.

Technology is driving this surge in climate adaptation. And that is why it is emphasized in this book. And this type will continue to surge for a long time. With AI-powered flood forecasting, open-source mapping tools, and community-controlled data platforms, non-state actors are increasingly able to assess risks, model scenarios, and organize responses without reliance on centralized expertise. Initiatives such as Google's Crisis Maps, Indigenous Climate Observatories, and AI-enabled Bioregional Adaptation Hubs enable communities to develop their adaptation blueprints from the ground up.

This reconfiguration challenges the dominant paradigm of climate governance. It does not reject the role of the state but insists that effective Adaptation must flow from the ground upward. In many cases, governments are now following the lead of these decentralized innovators, integrating community-designed plans into formal frameworks or co-governing adaptation strategies with civil society.

Climate change does not recognize borders. Its impacts—rising seas, shifting rainfall patterns, melting glaciers, migrating species, transboundary air pollution, wildfires, storm intensification, and climate-induced displacement—move across landscapes regardless of political boundaries. Changes in ocean currents can significantly impact storm intensification and the habitability of countries. Yet the dominant architecture of climate adaptation remains bound by nation-state logic, even as the crisis demands cross-border collaboration, bioregional planning, and post-sovereign thinking. As adaptation becomes more urgent, the tension between fixed territorial boundaries and fluid ecological systems is sharpening. These political boundaries have long been manipulated to hide human environmental impacts such as pollution. It has long been the practice to dump waste just over the nearby border. In the U.S., this waste is dumped on Indigenous Reservations in a process known as midnight dumping. Midnight dumping, also known as illegal dumping, is a common occurrence in many rural areas worldwide. Ocean dumping is a significant issue that has been growing with minimal oversight and control. Shifting ocean currents, rising sea levels, and intensification of storms transport this waste. Those that are easily detectable, such as radioactive hazardous wastes, have moved from the seas off the coast of Italy to North Africa, for example. Virtually all the coasts in the world are covered with the gigantic amount of plastic we have created.

The Ecological Reality vs. the Political Map

Earth's systems—rivers, watersheds, forests, mountain ranges, and atmospheric flows—are not confined to countries. AI and global communication make the Climate Adaptation generation painfully aware of this. Yet adaptation policies, funding, and legal mandates are overwhelmingly national. International adaptation finance flows through sovereign governments. National Adaptation Plans (NAPs), required under the UNFCCC, are structured by territorial jurisdictions. But floods don't stop at borders, nor do droughts, wildfires, intense storms, disease vectors, or climate migrants.

For example, poor resource management caused by political boundaries is evident in the Indus River basin, which spans India and Pakistan, both nuclear-armed rivals. Melting Himalayan glaciers threaten the water security of hundreds of millions, yet the political boundary inhibits joint watershed management. Similarly, the Sahel region faces desertification, migration, and conflict across Mali, Burkina Faso, Niger, and Nigeria—but adaptation programs remain fragmented and state-centric if they exist at all.

Bioregional Adaptation: A Border-Transcending Model

Some regions are beginning to adapt through bioregional cooperation, recognizing that ecosystems—not political lines—should guide resilience planning. The Mekong River Commission, involving Cambodia, Laos, Thailand, and Vietnam, coordinates flood management, fisheries, and water flows at a regional scale. In East Africa, IGAD (Intergovernmental Authority on Development) supports cross-border drought early warning systems among pastoralist communities that traditionally migrate in response to climate variability.

These examples show that transboundary governance can overcome narrow national interests. However, many such efforts remain underfunded or politically constrained, especially when extractive industries, militarized borders, or nationalist politics are involved.

Borders, Adaptation Finance, and Global Inequality

Country boundaries also mediate who receives adaptation finance and how. Wealthier nations and formal states are better positioned to secure funds from the Green Climate Fund (GCF) or World Bank. Stateless peoples, Indigenous nations, and border communities often fall outside eligibility—despite being among the most affected.

The result is a border-shaped injustice: adaptation funds flow along lines of recognition and state capacity, not need. Moreover, adaptation is often channeled into state-centric infrastructure (such as dikes, seawalls, and dams) that may reinforce elite control rather than support borderland communities, refugees, or ecological restoration.

Militarized Adaptation: Borders as Barriers, Not Bridges

Some governments are responding to climate instability with border militarization, treating adaptation as a matter of national security. In the U.S., DHS has integrated climate risk into border enforcement. European states are investing in high-tech fences and sea patrols while defunding humanitarian rescue efforts. Adaptation, in these cases, becomes a strategy of exclusion—fortressing the privileged while abandoning the displaced.

This model represents "climate apartheid"—a world where national boundaries are hardened to protect the global North. In contrast, the global South bears the brunt of displacement, food insecurity, and ecological collapse.

Examples of New Ways of Thinking about Systemic Approaches to Climate Change Adaptation

1. Cuba's Agroecological Transformation (Post-Soviet Era to Present)

Following the collapse of the Soviet Union in the early 1990s, Cuba experienced a sudden decline in oil, fertilizer, and trade support. Rather than succumbing to capitalist restructuring, it reorganized its food system around socialist principles of urban agriculture, food sovereignty, and ecological farming. The state redistributed land to worker cooperatives, promoted polyculture and permaculture techniques, and supported agroecological education through public institutions.

Today, Cuba is a globally recognized model for climate-resilient agriculture in a low-emissions, post-oil economy.

2. Zapatista Autonomous Zones in Chiapas, Mexico

Since 1994, the Zapatista movement has established autonomous Indigenous communities in southern Mexico that practice collective land ownership, democratic governance, and sustainable farming outside of state and corporate control. These communities, organized into caracoles (regional councils), resist capitalist development megaprojects, such as dams, highways, and monoculture plantations.

Zapatista adaptation includes reforestation, seed sovereignty, and Indigenous knowledge systems that enhance ecological resilience to droughts and storms in the region. They do not use the language of "adaptation"—but their practices embody a Marxist, anti-capitalist framework of ecological autonomy.

3. South Africa's Climate Justice Charter Movement

In post-apartheid South Africa, capitalist extraction continues to drive water inequality, land degradation, and energy injustice. In response, the Climate Justice Charter Movement, led by trade unions, food growers, township activists, and socialist thinkers, has launched a radical blueprint for climate adaptation rooted in decolonization and the principles of the food and commons.

They demand democratic control of water systems, climate reparations, and the end of fossil fuel capitalism—rejecting World Bank-style green finance in favor of public trust, land, and community food gardens.

4. Rojava (Northeast Syria): Democratic Confederalism and Ecological Restoration

Amid war and state collapse, the autonomous region of Rojava has implemented a system of democratic confederalism inspired by the ideas of Murray Bookchin and Abdullah Öcalan. This system integrates gender equality, ecological regeneration, and communal ownership of resources.

Despite siege and war, Rojava's communities are restoring rivers, resisting desertification, and reviving traditional farming under cooperative governance.

5. Landless Workers' Movement (MST) in Brazil

The Movimento dos Trabalhadores Rurais Sem Terra (MST) is Latin America's most significant social movement, organizing over one million people through land occupations, cooperative farming, and agroecological training centers. MST's farms serve as adaptation zones, planting native crops, restoring degraded land, and creating seed banks free from Monsanto's genetically modified (GM) patents.

MST explicitly frames climate resilience as a class and land struggle, calling for structural transformation and agrarian socialism to confront ecological collapse and rural poverty.

Federal agencies monitor global hotspots of climate-induced migration, integrate AI-driven border enforcement, and develop contingency plans to secure energy and food systems against external shocks. In Europe, nations invest in surveillance drones and maritime patrols while defunding refugee rescue operations.

Adaptation, in this context, becomes a mechanism of exclusion—a strategy to protect the few while fortifying against the many. Scholars have described this as "climate apartheid," a world in which the wealthy shield themselves through walls and technology while the poor face the full brunt of environmental collapse.

Despite these trends, there are growing movements to imagine and implement post-sovereign adaptation. This does not mean abandoning all forms of national governance but rather recognizing that effective climate adaptation will require bold new frameworks of cooperation, mobility, and shared responsibility. Bioregional governance rooted in watersheds, food systems, and migratory corridors offers one pathway. Another is the creation of climate mobility agreements that expand protections and legal pathways for displaced populations. Indigenous-led adaptation plans that transcend colonial borders—such as those by Sámi communities in the Arctic or Andean nations in South America—demonstrate how climate action can be grounded in ecosystems and cultural continuity rather than national claims.

Finally, post-sovereign adaptation demands a new ethic: one that replaces competition with cooperation and national exceptionalism with planetary solidarity. As the Doughnut Economics Action Lab argued in its 2025 Global Adaptation Charter, "The future of climate adaptation will be defined not by the map we inherited, but by the solidarity we create." This future cannot be built within the narrow confines of borders. It must emerge through new alignments of people, land, and knowledge that reflect the reality of a shared and interdependent planet.

The Digital Divide and Power for Global Climate Adaptation

Globally, the digital divide by age is one of the most persistent and consequential forms of inequality in the digital era. While young people are generally well-connected and actively engaged online, older adults are often left behind, creating disparities not only in access but also in their ability to participate fully in economic, health, and civic life. Due to the political influence of older individuals, as discussed in Chapter One, the digital divide poses a significant barrier to climate adaptation.

Young people aged 15 to 24 are the most connected demographic worldwide, with approximately 79% of individuals in this age group using the internet.

This figure significantly surpasses the global average of around 65% for the rest of the population. In regions such as Africa, the gap is even more pronounced: 55% of youth are online, compared to only 36% of older adults. This trend reflects a generational shift toward digital fluency among those who grew up with smartphones, social media, and on-demand digital services.

In contrast, older adults, especially those aged 65 and older—continue to face significant barriers to digital inclusion. In high-income and powerful countries, such as the United States, while internet use is nearly universal among adults aged 18 to 64, only about 75% of those aged 65 and above go online. In countries like Brazil, where climate impacts are likely, the situation is starker: nearly half of people over 60 report never having used the internet. These patterns are consistent across both developed and developing regions, often reflecting a combination of access, literacy, and attitudinal challenges.

Several underlying factors contribute to this age-based digital divide. The most fundamental is the generational difference in exposure to technology. Those born before the internet revolution often lack foundational digital skills and confidence, making them less likely to engage online. Cost and infrastructure also play a role, particularly in rural or economically disadvantaged communities where many older individuals reside. Health-related barriers—such as vision loss, reduced dexterity, or cognitive decline—can make the use of modern devices more difficult. Finally, cultural and psychological factors come into play. Some older individuals may perceive technology as unnecessary, intimidating, or even dangerous, especially if they lack strong social support for learning.

The implications of this divide are profound. Perhaps most concerning, the lack of access to timely, trustworthy health information, especially during global health emergencies—can directly affect their well-being and safety.

Efforts to close the digital divide by age are increasingly prioritized in both international and national development strategies. Initiatives such as the UN's Global Digital Compact and the World Economic Forum's Edison Alliance aim to ensure digital equity through partnerships between governments, private companies, and civil society. Community-based programs in countries such as New Zealand and the United Kingdom have made significant strides in promoting digital literacy, providing devices to low-income seniors, and designing age-friendly digital interfaces.

Organizations such as the Good Things Foundation have shown how targeted outreach can improve confidence, skills, and access among older adults.

The digital divide by age reflects a broad and persistent gap between younger and older populations, one that has direct consequences for equity, participation, and well-being in an increasingly digital world. Closing this divide requires not just infrastructure but human-centered approaches—policies and programs that recognize the specific needs and strengths of older adults while supporting intergenerational learning and inclusion. As society ages and technology evolves, bridging this gap will be essential for a just and flourishing digital future.

The use of Artificial Intelligence (AI) in climate change adaptation holds transformative potential. Still, its benefits are unequally distributed due to the global digital divide—particularly across age, income, and geographic lines. AI technologies can help predict climate risks, support disaster response efforts, and inform long-term planning for resilience. Still, their effectiveness is limited if populations, especially older people, rural communities, and people without digital access—cannot engage with or benefit from these tools.

D. AI and Climate Adaptation Futures

AI is increasingly used to enhance climate forecasting, simulate future environmental scenarios, and optimize adaptive infrastructure. Machine learning models can analyze vast amounts of data—from satellite imagery to sensor networks—to detect early warning signs of droughts, floods, wildfires, and heatwaves. AI tools also support precision agriculture, enabling farmers to adjust their planting strategies in response to evolving climate conditions. In cities, AI helps optimize energy systems, model urban heat islands, and design infrastructure to withstand extreme weather events. For governments and emergency services, real-time AI systems can process information from social media, drones, and sensor networks to improve the speed and precision of crisis responses.

However, these advancements presuppose access to digital infrastructure, basic digital literacy, and the capacity to interact with AI-powered tools. The global digital divide, particularly by age, creates significant barriers to climate resilience. Older adults, especially in low- and middle-income countries, are significantly less likely to access the internet or use smartphones, which are often the platforms for receiving AI-generated early warnings, health advisories, or adaptation recommendations.

Without access to these tools, older populations may miss evacuation alerts, heat wave warnings, or water conservation instructions generated by AI systems.

This technological exclusion compounds existing vulnerabilities. Older adults are often more physically susceptible to heat stress, extreme weather, and health crises. If they are digitally disconnected, they are less likely to benefit from telemedicine during heatwaves, app-based ride services during evacuations, or digital government portals distributing emergency aid. The same applies to rural populations, many of whom may be disconnected not only from the internet but also from the AI-driven climate services being rolled out in urban and wealthier areas. This inequity risks deepening the divide between those who can adapt and those left behind.

Moreover, as AI becomes a core tool in adaptation planning, there is a growing risk that the perspectives and needs of digitally marginalized groups, especially older adults, are underrepresented in datasets and decision-making processes. AI models trained primarily on data from connected, tech-savvy, younger urban populations may produce results that overlook the realities of older, rural, or Indigenous communities. This exclusion not only weakens the effectiveness of adaptation strategies but also erodes trust in climate governance and public institutions.

To address this issue, adaptation policies must incorporate digital inclusion as a foundational principle. This includes investing in internet access for underserved areas, offering age-appropriate digital literacy programs, and designing AI tools with user-friendly interfaces for all ages. Community-based intermediaries—such as libraries, community centers, and health clinics—can serve as trusted access points for AI-generated information, bridging the gap for populations who are not digitally native. Policymakers must also mandate inclusive data collection and human oversight of AI tools to ensure that algorithms reflect the full diversity of climate vulnerability and lived experience.

In conclusion, while AI has the power to revolutionize climate adaptation, its success depends on addressing the persistent and growing digital divide. Without equitable access to AI-enabled tools and services, the very communities most at risk from climate change, especially older adults, rural residents, and the digitally excluded—will continue to face disproportionate harm. Bridging the digital divide is not only a matter of technological fairness; it is a necessary step for building a just and climate-resilient future.

E. Cumulative Impacts of Climate Change

Cumulative climate change impacts refer to the combined and compounding effects of multiple climate-related stressors over time and across sectors. Unlike single-event disasters, cumulative impacts build gradually, interact with one another, and often amplify existing social, environmental, and health vulnerabilities—especially in marginalized or underserved communities.

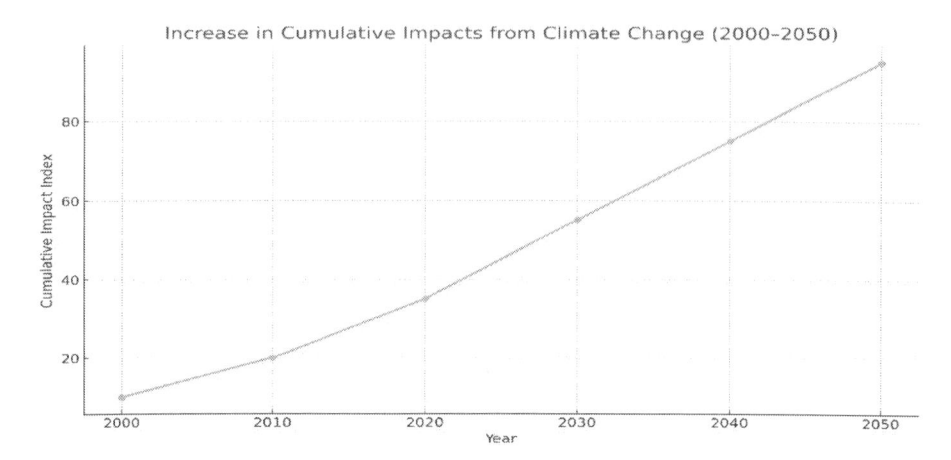

Increase in Cumulative Impacts from Climate Change (2000-2050)

Understanding Cumulative Impacts

Cumulative climate impacts are not isolated occurrences. They unfold through the interaction of multiple hazards, including sea-level rise, heat waves, drought, wildfires, extreme rainfall, and biodiversity loss. These effects often accumulate over decades and may lead to irreversible changes in ecosystems, infrastructure, food systems, and public health. For example, repeated heatwaves can exacerbate drought conditions, reduce agricultural yields, intensify wildfire seasons, and contribute to long-term water insecurity.

In urban areas, rising temperatures, poor air quality, and increased flooding may strain aging infrastructure, causing disproportionate harm to low-income communities. In coastal zones, sea-level rise, saltwater intrusion, and erosion interact with socioeconomic vulnerabilities, making recovery from each event harder over time. These cumulative effects are particularly harsh for people who face overlapping disadvantages—such as older people, Indigenous populations, and those living in poverty.

Examples of Cumulative Climate Impacts

In California, years of drought followed by intense rainfall events have compromised water infrastructure, leading to both agricultural collapse and flash flooding—two contrasting yet interconnected hazards. In the Arctic, permafrost thaw, sea ice loss, and coastal erosion are simultaneously displacing Indigenous communities, destabilizing housing, and threatening traditional ways of life. In small island developing states, the cumulative pressure of sea-level

Cumulative Impacts on Health and Well-being

Cumulative climate effects also pose long-term risks to public health. Chronic exposure to heat increases the risk of cardiovascular disease and heat-related mortality. Air pollution from recurring wildfires can have a lasting impact on respiratory health. Prolonged droughts and floods degrade water quality, spread vector-borne diseases, and disrupt mental health through eco-anxiety, displacement, and trauma. Children and older adults are especially vulnerable due to their physiological sensitivity and limited adaptive capacity.

In Indigenous and frontline communities, cumulative impacts erode not only material resources but also cultural practices, food sovereignty, and ancestral connections to land. Economic damage assessments cannot fully capture these losses but are critical for understanding long-term resilience.

Why Cumulative Impacts Matter for Planning

Most climate risk assessments, insurance models, and government planning frameworks still focus on single hazards or short-term economic losses. However, failing to account for cumulative and compounding effects underestimates long-term damage and results in inadequate adaptation strategies. For example, a flood protection plan that ignores concurrent heat stress or drought may fail to address the full range of community vulnerabilities.

Effective climate adaptation must, therefore, include cumulative impact assessments that:

- Analyze long-term exposure to multiple hazards,
- Consider social and ecological feedback loops,
- Center the experiences of overburdened communities,
- Integrate public health, housing, and environmental justice into planning,
- Utilize scenario modeling and AI to forecast the combined effects of stressors.

Legal and Policy Relevance

Cumulative climate impacts are increasingly central to legal arguments and environmental justice movements. In the U.S., cumulative impact assessment is now being incorporated into some state-level environmental regulations (e.g., New Jersey's Environmental Justice Law). At the international level, climate-vulnerable nations are calling for loss and damage frameworks that reflect long-term, non-economic, and intergenerational harm.

Incorporating cumulative impacts into national adaptation plans, infrastructure design, insurance, and health policy is essential for building resilient communities, especially as the frequency and intensity of climate disruptions increase in a warming world.

F. Conclusion

The challenges of cross-boundary climate impacts, digital divides, AI, and cumulative effects will not disappear. Nor will climate change. Only by adapting to climate change bioregionally, incorporating inclusive technology, and accounting for cumulative impacts will the climate adaptation generation envision a flourishing future for everyone.

Resources

1. Cruz, M. J., & Kumar, S. (2024). Artificial intelligence in climate adaptation and disaster resilience. Climate Intelligence Review, 11(2), 45–61.
2. FEMA AI Futures Lab. (2025). Integrating machine learning in federal disaster planning. U.S. Department of Homeland Security.
3. Google Crisis AI Report. (2024). Flood forecasting and AI-powered alerts: Lessons from South Asia. Google Research.
4. IFRC Simulation & Tech Futures Report. (2025). Next-gen emergency responder training with AI and VR. International Federation of Red Cross and Red Crescent Societies.
5. Natural Resources Canada. (2024). AI-guided wildfire management in Canada: 2023 pilot summary. Government of Canada.
6. Munich Climate Risk Lab. (2024). AI and insurance: Redesigning risk in the era of climate volatility. Munich Re.
7. United Nations Environment Programme (UNEP). (2024). Harnessing AI for climate adaptation: Global guidelines and ethical frameworks. https://www.unep.org

Digital Divide and Climate Vulnerability (Especially by Age)

1. International Telecommunication Union (ITU). (2024). Measuring digital development: Facts and figures 2024. https://www.itu.int
2. Pew Research Center. (2024). Digital divide persists among older adults despite broadband growth. https://www.pewresearch.org
3. World Economic Forum. (2025). Bridging the digital divide in climate resilience: Equity, aging, and access. Geneva: WEF Digital Equity Taskforce.
4. UNDP. (2024). Digital inclusion for climate justice: Vulnerability mapping and AI access. United Nations Development Programme. https://www.undp.org
5. Good Things Foundation. (2024). Digital skills and resilience: Supporting older adults in a climate-impacted world. UK Digital Inclusion Report.
6. World Bank. (2025). Technology and adaptation: Ensuring AI reaches the most vulnerable. Washington, DC: Global Adaptation Initiative.
7. United Nations Environment Programme (UNEP). (2024, September). Artificial Intelligence end-to-end: The environmental impact of the full AI lifecycle needs to be comprehensively assessed. https://www.unep.org
8. UNEP. (2024, May). AI has an environmental problem. Here's what the world can do about that. https://www.unep.org/news-and-stories/story/ai-has-environmental-problem-heres-what-world-can-do-about
9. UNEP, ITU, & WMO. (2024). AI for environmental monitoring and climate adaptation: Tools and partnerships for resilience. United Nations Innovation Network. https://www.unep.org
10. United Nations Economic and Social Council (ECOSOC). (2024, May). Harnessing AI for climate resilience: Statements from the High-Level Political Forum. United Nations Press Release. https://press.un.org
11. United Nations Framework Convention on Climate Change (UNFCCC). (2024). Artificial Intelligence for Climate Action in Developing Countries. Technology Executive Committee Report. https://unfccc.int
12. United Nations University – Institute for Water, Environment and Health (UNU-INWEH). (2024). Harnessing the power of AI for water security and climate change impact assessment. United Nations Digital Library. https://digitallibrary.un.org
13. UNEP. (2022). Adaptation Gap Report 2022: Too Little, Too Slow – Climate adaptation failure puts world at risk. https://www.unep.org/resources/adaptation-gap-report-202.

Glossary of Key Terms

Adaptation
The process of adjusting to current or expected climate changes to reduce harm or take advantage of beneficial opportunities. Adaptation can be individual, community-wide, or systemic, including new policies, infrastructure, or behaviors.

Bioregion
A geographic area defined by natural boundaries such as watersheds, climate, and ecosystems rather than political lines. It emphasizes local ecological knowledge and sustainability practices rooted in place-based living.

Citizen Science
Public participation in scientific research, where individuals—especially students and communities—gather and analyze data related to environmental or climate issues. Examples include tracking bird migrations or recording local air quality.

Climate Anxiety
Feelings of fear, stress, or helplessness associated with climate change and its effects. Often experienced by youth and frontline communities, it underscores the need for emotional support and proactive coping strategies.

Climate Justice
A framework that connects climate change to social justice, emphasizing that those least responsible for emissions often suffer the most from climate impacts. It promotes fair treatment, reparative action, and inclusion of vulnerable groups in decision-making.

Climate Literacy
Understanding the science of climate change, its societal impacts, and the solutions available. It includes systems thinking, data analysis, and the ability to take informed action.

Cumulative Risk Assessment
An environmental assessment method that evaluates the combined risks of multiple pollutants, exposures, and stressors on human and ecological health—especially relevant for overburdened communities.

Ecological Resilience
The capacity of ecosystems to recover from disturbances like fires, floods, or human impacts, while maintaining essential functions and services.

Experiential Learning
Learning through direct, hands-on experience. In climate education, this includes gardening, field trips, simulations, and real-world problem solving to enhance engagement and understanding.

Flourishing
A concept from psychology and philosophy that refers to holistic well-being, including meaning, connection, purpose, and personal growth. In a climate context, it integrates emotional health, community resilience, and environmental stewardship.

Green Infrastructure
Nature-based systems designed to manage water, reduce heat, and improve urban living—such as green roofs, tree-lined streets, wetlands, and permeable surfaces.

Intergenerational Leadership
Collaboration between youth and older generations to share wisdom, build climate resilience, and shape policy. This model respects past experiences while empowering future visions.

Land Use
The management and modification of natural environments for agriculture, housing, transportation, recreation, industry, or conservation. Thoughtful land use planning helps balance development with environmental protection, reduces climate risk, and preserves biodiversity and local food systems.

Place-Based Education
An educational approach that uses the local environment and community as a starting point for learning, fostering a deeper connection to land, culture, and sustainability practices.

Resilience
The ability of individuals, communities, or systems to anticipate, prepare for, respond to, and recover from climate-related disruptions while maintaining core functions.

Trauma-Informed Education
An educational approach that acknowledges how trauma—including from climate disasters—affects learning, and seeks to create safe, supportive, and empowering learning environments.

Urbanization
The ongoing process of population growth in cities and metropolitan areas. While urbanization can strain infrastructure and contribute to emissions, it also offers opportunities for innovation in sustainable design, transportation, housing, and public health. Managing urban growth is essential for equitable and climate-resilient development.

Visioning
A process of imagining and articulating positive, just climate futures. Visioning is used in community planning, education, and policy design to inspire hope and guide action.

Youth Climate Advocacy
Young people leading movements to influence climate policy, raise awareness, and demand accountability. Includes protests, policy proposals, peer education, and climate organizing.

Zero-Carbon Solutions
Technologies and strategies that produce no net greenhouse gas emissions—such as renewable energy, passive building design, and regenerative agriculture.

About the Author

I have taught, researched and published in law, social work, and urban and environmental planning for over 40 years. I have earned graduate degrees in all three fields. I have also been appointed to advisory commissions by 3 state governors and 3 presidents. I have published many law reviews, 2 were the first on the topic in that genre. I have published 9 volumes of encyclopedias, some with my wife Robin Morris Collin. 2 were the first in that genre - the book on the EPA was mine. The 3-volume set on sustainability was ours. My book, Battleground Environment, 2 volumes, described the 100 most controversial environmental issues from the perspective of community, government, and industry. I have also been certified as an expert witness in US federal court and in state Court. I have won 2 awards teaching decided by students. The American Institute of Architects awarded an interdisciplinary committee I chaired to us for creating a new required course called "Environmental Choices". My recent book, Climate Change in the Classroom: Celebrating Optimism, is now available.
https://amazon.com/author/collin.com

Made in the USA
Columbia, SC
11 July 2025